Ultimate SEO
Grow Your Business Online

Lisa Anders

Publisher's Note

Every possible effort has been made to ensure that the information contained in this book is accurate at the time of publication. However, the publisher and author cannot accept responsibility for any errors or omissions, however caused.

Disclaimer

The information provided in this book series is intended for educational and informational purposes only. The advice and strategies contained herein are based on the author's extensive experience and research and are provided as general guidance.

All examples and case studies mentioned in this series are fictional and are used to illustrate concepts and strategies. They are not based on real-life situations or companies. The results described in these examples are hypothetical and may not reflect actual outcomes.

While every effort has been made to ensure the accuracy and completeness of the information, the author and publisher make no representations or warranties regarding its applicability to your specific situation. Readers are encouraged to seek professional advice tailored to their unique needs and circumstances. The author is available for professional consultation to provide customized advice and strategies.

The author and publisher shall not be held liable for any loss or damage, including but not limited to indirect or consequential loss or damage, arising out of or in connection with the use of the information contained herein.

eBook ISPN: 979-8-9910096-0-7
Paperback ISPN: 979-8-9910096-1-4
Hardback ISPN: 979-8-9910096-2-1

Contents

About The Author

Lisa Anders is a seasoned SEO specialist and digital marketing expert with nearly two decades of hands-on experience in the field. Her journey in the world of SEO began long before the term became a buzzword, and over the years, she has witnessed and adapted to the ever-evolving landscape of search engine optimization and digital marketing.

Lisa's expertise lies in her ability to demystify complex SEO concepts and translate them into practical, actionable strategies that deliver measurable results. Her career has been marked by a commitment to helping businesses of all sizes harness the power of SEO to achieve their online goals. From startups to established enterprises, Lisa has worked with a diverse range of clients, each time tailoring her approach to meet their unique needs and challenges.

In addition to her extensive professional experience, Lisa is also a passionate educator and mentor. She has conducted numerous workshops and training sessions, empowering business owners and marketing professionals with the knowledge and skills needed to thrive in the digital age. Her teaching style is characterized by clarity, enthusiasm, and a genuine desire to see others succeed.

Lisa's dedication to excellence and her results-driven approach have earned her a reputation as a trusted advisor in the SEO community. She is known for her no-nonsense, straightforward advice that cuts through the noise and focuses on what truly matters – achieving sustainable growth and success.

Ultimate SEO: Grow Your Business Online is a culmination of Lisa's years of experience, distilled into a comprehensive guide that aims to equip readers with the tools and strategies necessary to excel in the competitive online marketplace. Through this book, she hopes to share her knowledge and passion for SEO, helping readers unlock their business's full potential.

When she's not diving into the latest SEO trends and techniques, Lisa enjoys spending time with her family, exploring new technologies, and contributing to online forums and communities where she continues to share her insights and learn from others.

Lisa Anders is more than just an SEO expert; she is a dedicated advocate for digital excellence and a mentor to those looking to make their mark online.

Take Your SEO to the Next Level!

Thank you for reading Ultimate SEO: Grow Your Business Online! Your journey to mastering SEO and digital marketing has just begun, and there are many more ways you can continue to enhance your skills and grow your business.

Stay Connected

Join our community of like-minded entrepreneurs and marketers who are committed to achieving online success. Here's how you can stay connected and get the most out of your SEO journey:

1. Sign Up for Exclusive Updates and Tips

Stay ahead of the curve with the latest SEO trends, tips, and updates. Visit www.lisanders.com to sign up for our newsletter and receive exclusive insights, updates on new tools and resources, invitations to webinars and live Q&A sessions, and special offers and discounts on additional resources.

2. Share Your Thoughts

Your feedback is invaluable. If you found this book helpful or have suggestions for improvement, I'd love to hear from you. Sharing your experiences helps me provide even more valuable content in the future.

3. Ask Questions

Got questions about SEO or digital marketing strategies? Feel free to reach out! I'm here to help you on your journey to online success. Whether it's a quick question or a detailed inquiry, don't hesitate to get in touch.

4. Explore More Resources

Looking for more in-depth knowledge? Explore additional resources, including advanced SEO courses, webinars, and consulting services designed to help you achieve your business goals.

Ready to Transform Your Business?

Start implementing the strategies you've learned today. Your journey to online success is in your hands. Let's make it happen! Visit www.lisanders.com for all the details.

Chapter 1

Introduction

What if I told you that mastering three key components of digital marketing could skyrocket your business growth and put you miles ahead of your competition? Sounds like a bold claim, doesn't it?

This book is your cheat code to the complex game of digital marketing. Much like understanding the hidden mechanics in a video game or a blockbuster film's plot twist, mastering digital marketing can transform your business's online presence and profitability.

Digital marketing isn't just about having a website or running ads; it's about strategically using these tools to attract, engage, and convert your audience. In the following chapters, we'll dive deep into each element, providing actionable insights and real-world examples to guide you every step of the way.

From boosting your search engine rankings with SEO (Search Engine Optimization), to driving targeted traffic through PPC (Pay-Per-Click or Paid Advertising), to creating a user-friendly website design that converts visitors into customers—this book covers it all.

While many think of these elements as separate strategies, they are most powerful when integrated into a cohesive digital marketing plan. By the end of this book, you'll not only understand each component but also how to weave them together to create a comprehensive strategy that delivers actual results.

Prepare to embark on a journey that will change the way you view and implement digital marketing. Whether you're a seasoned business owner or just starting out, this guide is your key to unlocking unprecedented growth and profitability.

Ready to dive in? Let's get started.

Chapter 2

UNDERSTANDING YOUR IDEAL CUSTOMER

Imagine having a map that shows exactly who your customers are, what they want, and how you can reach them. Sounds like a game-changer, right? That's the power of understanding your ideal customer. Knowing your ideal customer is the foundation of any successful marketing strategy. It's about understanding who they are, what they need, and how your product or service can solve their problems.

Working with various businesses, I've often asked the question, "Who is your ideal customer?" Surprisingly, many respond with, "Everyone." They believe they don't have a specific target audience. If this sounds familiar to you, my advice is straightforward: marketing to everyone often means you're spreading yourself too thin.

Your messaging becomes too general, and as a result, you may not truly resonate with anyone. Instead, focus on identifying and targeting your ideal customers. This approach allows you to create more personalized, compelling messages that can significantly improve your marketing effectiveness.

Digital marketing isn't just about broadcasting your message to a wide audience; it's about connecting with the right people in a meaningful way. This chapter will guide you through the steps of conducting audience research, creating customer personas, and using these insights to inform your marketing strategy.

By the end of this chapter, you'll have a clear understanding of how to determine who your ideal customers are and how to reach them effectively. Whether you're a seasoned business owner or a startup, mastering the art of understanding your ideal customer is the foundation for building a successful marketing strategy that drives growth.

Ready to discover who your ideal customers are and how to connect with them? Let's dive into the process of understanding your ideal customer and transforming your marketing strategy.

Steps to Define Your Ideal Customer

- **Conduct Audience Research:**
  - o **Surveys:** Use online survey tools to gather data on demographics, interests, and behaviors.
  - o **Interviews:** Conduct one-on-one interviews with your current customers to gain deeper insights.

Many marketing companies skip this step, fearing that directly reaching out to customers might be intrusive. While direct research is immensely valuable, it can be challenging. If direct interaction isn't possible, there are alternative methods to define your ideal customer.

  - o **Social Media Analysis:** Analyze your social media followers to understand their interests and engagement patterns.
- **Analyze the Data:**
  - o **Google Analytics:** Use this tool to identify common characteristics and patterns in your website visitors, such as age, location, and interests.
  - o **Customer Relationship Management (CRM) Systems:** Leverage CRM data to find trends and behaviors among your existing customers.
- **Create Customer Personas:**
  - o **Demographics:** Age, gender, income, education level, occupation.
  - o **Psychographics:** Interests, values, lifestyle, personality traits.

- Goals: What they aim to achieve with your product or service.
    - Pain Points: Challenges or problems they face that your product or service can solve.
- **Use Customer Insights:**
    - **Tailor Marketing Messages:** Craft messages that directly address the needs and pain points of your customer personas.
    - **Personalize Content:** Create personalized content and campaigns that resonate with different segments of your audience.
- **Segment Your Audience:**
    - **Group Customers:** Based on similar characteristics such as demographics or behavior.
    - **Targeted Campaigns:** Develop specific campaigns for each segment to maximize engagement and conversion.

Example: Analyzing Data to Create Customer Personas

Data Analysis Example for a Service-Based Business:

Let's say you run a local home cleaning service. Here's how you might analyze your data and use it to create a customer persona:

- **Google Analytics Data:**
    - **Demographics:** Majority of visitors are aged 35-50, 70% female.
    - **Geography:** High traffic from suburban areas within a 30-mile radius.
    - **Behavior:** Visitors often explore the service packages page and the testimonials but leave without booking.
- **Survey Results:**
    - **Interests:** High interest in eco-friendly cleaning products and services.

- **Pain Points:** Lack of time to clean due to busy schedules, concerns about the use of harsh chemicals.

Creating Customer Personas

Persona #1: Busy Brenda

- **Age:** 40
- **Gender:** Female
- **Location:** Suburban area
- **Occupation:** Marketing Executive
- **Interests:** Spending time with family, eco-friendly products, maintaining a clean and healthy home
- **Goals:** Wants a reliable and thorough cleaning service that uses eco-friendly products.
- **Pain Points:** Struggles with balancing a busy work schedule and maintaining a clean home without using harsh chemicals.

Persona #2: Senior Sam

- **Age:** 65
- **Gender:** Male
- **Location:** Suburban area
- **Occupation:** Retired
- **Interests:** Gardening, spending time with grandchildren, health and wellness
- **Goals:** Needs assistance with house cleaning to maintain independence and a clean living environment.
- **Pain Points:** Difficulty with physically demanding cleaning tasks, concerns about affordability and reliability of the service.

Tailored Marketing Messages

For Busy Brenda:

"Meet Busy Brenda, a dedicated marketing executive juggling a hectic work schedule and family time. She needs a home cleaning service that's thorough, reliable, and eco-friendly. Our cleaning services use non-toxic, eco-friendly products that are safe for her family and the environment. Let us take care of the cleaning, so Brenda can focus on what matters most. Book your eco-friendly cleaning service today and enjoy a spotless home without the hassle."

For Senior Sam:

"Introducing Senior Sam, a retired gentleman who enjoys gardening and spending time with his grandchildren. He needs a dependable cleaning service to help maintain his home's cleanliness and his independence. Our team provides thorough cleaning services tailored to Sam's needs, ensuring a healthy living environment. Plus, our affordable packages make it easy for seniors to keep their homes spotless. Schedule a cleaning today and let us help you maintain your home with care and respect."

Utilizing Your Customer Insights

Once you have developed a clear understanding of your ideal customers and crafted tailored marketing messages, the next step is to leverage this information across various marketing initiatives. Here's how you can use these insights to create effective web pages, blog posts, and targeted Google Ads campaigns:

Creating Web Pages:

- **Landing Pages:** Design dedicated landing pages for each customer persona, addressing their specific needs and pain points.

- **Service Pages:** Highlight the benefits of your services in a way that resonates with each persona. For example, a service page for eco-friendly cleaning can be tailored to appeal to Busy Brenda's concerns about harsh chemicals.
- **Testimonials and Case Studies:** Use testimonials that reflect the experiences and satisfaction of customers similar to your personas. This builds trust and demonstrates your understanding of their needs.

Writing Blog Posts:

- **Content Topics:** Choose blog topics that address the interests and pain points of your personas. For example, a blog post titled "5 Eco-Friendly Cleaning Tips for Busy Professionals" would be highly relevant to Busy Brenda.
- **SEO Optimization:** Use keywords and phrases that your personas are likely to search for. This improves your content's visibility and attracts the right audience.
- **Engagement:** Encourage engagement by asking questions and inviting feedback that resonates with your personas' experiences and concerns.

Developing Google Ads Campaigns:

- **Targeted Ads:** Create ad campaigns specifically tailored to each persona. Use the language and messaging that speaks directly to their needs and goals.
- **Ad Extensions:** Utilize ad extensions to provide additional information that might appeal to your personas, such as eco-friendly certifications for Busy Brenda or senior discounts for Senior Sam.
- **A/B Testing:** Conduct A/B testing with different ad copies to see which messages resonate best with each persona, refining your approach based on the results.

Persona Service Page Targeting

Creating a service page that targets your customer persona is essential for building relevance and engagement. When potential customers find content that directly addresses their specific needs and pain points, they are more likely to engage and consider your services. For instance, a service page for eco-friendly cleaning services tailored to busy professionals will resonate because it shows that your company understands their need for efficient and environmentally safe solutions.

Tailoring your service page to a customer persona personalizes the user experience, fostering trust and increasing the likelihood of converting visitors into clients. A service page designed for seniors that emphasizes reliability and affordability will connect more deeply with this audience than a generic page.

Targeted messaging improves conversion rates by allowing potential customers to see the direct benefits of your service, making them more likely to take action. Highlighting flexible scheduling for busy professionals or affordable packages for seniors makes your service more appealing and actionable.

Along with improving conversions/lead generation creating service pages that target specific customer personas can improve your SEO performance. By incorporating keywords and phrases that your target audience is likely to use when searching for solutions, you increase the chances of your page ranking higher in search engine results.

This targeted approach can attract more organic traffic from users who are specifically looking for the services you offer. For example, using keywords like "eco-friendly cleaning for professionals" or "senior-friendly home cleaning" can help your page rank better for those search queries.

A targeted service page helps differentiate your business from competitors by addressing unique needs and providing tailored solutions. This specialization sets you apart from generic providers, making your business the preferred choice for your target customers.

Finally, a targeted service page allows you to clearly communicate your value proposition. By focusing on what matters most to your customer persona, you highlight the unique benefits of your service, helping potential customers quickly understand why they should choose you.

Persona Service Page Example

For Busy Brenda:

- **Heading Section:**
    - Home Cleaning Services for Busy Professionals
        - Your Trusted Partner in Eco-Friendly Home Cleaning
        - Are you a busy professional struggling to find time to keep your home clean? Do you worry about the harmful effects of harsh cleaning chemicals on your family and the environment? At [Your Company Name], we understand the unique challenges faced by professionals like you, and we are here to help.
        - Call [Phone Number]
- **Next Section:**
    - Why Choose Our Eco-Friendly Cleaning Services?
        - Time-Saving Convenience
        - Flexible Scheduling: We offer flexible scheduling options to fit your busy lifestyle. Whether you need weekly, bi-weekly, or monthly cleanings, we can accommodate your needs.

- Quick and Efficient: Our professional cleaning team works quickly and efficiently, ensuring your home is spotless without disrupting your schedule.

For Senior Sam:

- **Heading Section:**
  - Home Cleaning Services for Seniors
    - Your Dependable Partner in Senior Home Cleaning
    - Are you a senior looking for assistance with home cleaning tasks? Do you want to maintain your independence while ensuring your home is clean and safe? At [Your Company Name], we specialize in providing thorough and affordable home cleaning services tailored to the needs of seniors.
    - Call [Phone Number]
- **Next Section:**
  - Why Choose Our Senior-Friendly Cleaning Services?
    - Reliable and Compassionate Service
    - Trusted Professionals: Our cleaning staff is experienced, background-checked, and trained to provide reliable and compassionate service. We treat your home with the respect and care it deserves.
    - Senior Discounts: Take advantage of our special rates for seniors to make our services even more accessible.

Persona Blog Content Targeting

For Busy Brenda: The blog topic "10 Tips for Maintaining a Clean Home with a Busy Schedule" directly addresses the pain points of busy professionals who struggle to keep their homes clean.

By providing practical tips and highlighting eco-friendly cleaning solutions, this content positions the business as a helpful resource. It showcases the company's understanding of the target audience's challenges and offers valuable advice.

This builds trust and credibility, making it more likely that busy professionals will consider the cleaning service when they need help.

For Senior Sam: The blog topic "How to Keep Your Home Clean and Safe in Your Golden Years" focuses on the specific needs of seniors who may find cleaning physically challenging.

By offering simple and effective cleaning routines and emphasizing the health benefits of a clean home, the content demonstrates empathy and understanding of the audience's situation. It also introduces the business's senior-friendly services, positioning the company as a caring and reliable partner.

This approach not only attracts seniors looking for cleaning assistance but also appeals to their family members who may be seeking solutions for their elderly loved ones.

Persona Google Ads Examples

We will explore a solid PPC strategy in greater detail in a later chapter, so we won't go in-depth here. These examples are intended to demonstrate how understanding your target customer can enhance your overall marketing strategy, **extending beyond SEO.**

For Busy Brenda:

- **Headline:** "Home Cleaning for Busy Professionals"
- **Description:** "No time to clean? Let us handle it with our eco-friendly products. Book now and get a sparkling home without the stress!"

- **Call to Action:** "Book Now"
- **Imagery:** A busy professional woman smiling with relief as she walks into a clean home.

This ad speaks directly to busy professionals who value eco-friendly products and struggle to find time for cleaning.

By highlighting the use of eco-friendly products and emphasizing stress-free cleaning, the ad addresses Brenda's pain points and demonstrates the company's expertise and reliability. This positions the business as a knowledgeable and trustworthy service provider that understands and can solve Brenda's specific needs.

For Senior Sam:

- **Headline:** "Affordable Home Cleaning for Seniors"
- **Description:** "Reliable and thorough cleaning services tailored for seniors. Maintain your independence and a clean home. Schedule today!"
- **Call to Action:** "Schedule Today"
- **Imagery:** An elderly man comfortably sitting in a clean and organized living room.

This ad targets seniors by emphasizing affordability and reliability, key concerns for this demographic. It reassures Sam that the service will help maintain his independence while keeping his home clean.

By showcasing the business's understanding of senior needs and offering tailored solutions, the ad positions the company as a trusted provider capable of delivering reliable and compassionate service.

Competitive Advantage: Creating highly relevant ads tailored to specific personas helps differentiate your business from competitors. By addressing the unique needs and concerns of your target audience, your ads stand out and capture the attention of potential customers more effectively than generic ads.

Persona Social Media Ad Examples

For Busy Brenda:

- **Ad Copy:**
    - **Text:** "Struggling to keep your home clean with a busy schedule? Our eco-friendly cleaning services are here to help. Enjoy more family time and a spotless home. Book now and get a special discount!"
    - **Call to Action:** "Learn More"
    - **Imagery:** A carousel ad featuring images of a clean kitchen, living room, and happy family time, highlighting the eco-friendly aspect.

For Senior Sam:

- **Ad Copy:**
    - **Text:** "Need help keeping your home clean? Our affordable cleaning services are perfect for seniors. Enjoy a healthy, clean living space without the hassle. Contact us today for special senior rates!"
    - **Call to Action:** "Contact Us"
    - **Imagery:** A warm, inviting image of an elderly man with his grandchildren in a clean and cozy home setting.

Persona Social Media Post Examples

For Busy Brenda:

- **Text:** "Struggling to keep your home clean with a busy schedule? Our eco-friendly cleaning services are here to help. Enjoy more family time and a spotless home. Book now and get a special discount!"
- **Call to Action:** "Learn More"
- **Imagery:** A carousel ad featuring images of a clean kitchen, living room, and happy family time, highlighting the eco-friendly aspect.

For Senior Sam:

- **Text:** "Need help keeping your home clean? Our affordable cleaning services are perfect for seniors. Enjoy a healthy, clean living space without the hassle. Contact us today for special senior rates!"
- **Call to Action:** "Contact Us"
- **Imagery:** A warm, inviting image of an elderly man with his grandchildren in a clean and cozy home setting.

Chapter 3

ANALYZING YOUR COMPETITORS

What if I told you that your competitors could be your greatest teachers? By studying their strategies, you can uncover valuable insights and opportunities for your own business. This isn't about copying what they do; it's about learning from their successes and mistakes, identifying gaps in the market, and discovering ways to differentiate your business.

Digital marketing success is not achieved in isolation. It requires a keen understanding of who you are up against and what tactics they are employing. We'll explore how to assess your competitors' SEO strategies, content quality, user experience, and promotional tactics.

Steps to Conduct a Competitive Analysis

- **Identify Your Competitors:**
  - List direct and indirect competitors.
  - Use market research tools to find key players in your industry.
- **Analyze Their Online Presence:**
  - Evaluate their website design and user experience.
  - Check their social media engagement and content.
- **Evaluate Their SEO Strategies:**
  - Analyze their keyword rankings and backlink profiles using tools like Ahrefs and SEMrush.
- **Assess Their Content:**
  - Review their blog posts, articles, and multimedia content.
  - Identify the type of content that resonates with their audience.
- **Understand Their Offers:**

- o Look at their products, services, pricing, and promotions.
- o Note any unique selling propositions (USPs) they highlight.

A thorough competitive analysis provides you with a strategic advantage, allowing you to identify gaps in the market, understand industry trends, and refine your own marketing strategies. This insight can help you make informed decisions that propel your business forward.

Examples of Competitor Analysis Insights

Targeting Industry-Specific Needs:

- **Resource Library:** Suppose you notice that a competitor's website has an extensive resource section dedicated to industry-specific case studies and whitepapers. Knowing that your target audience values detailed, relevant content, you could create a similar resource library on your site. This would position your business as an expert in the field and attract industry professionals looking for reliable information.
- **Dedicated Industry Pages:** If you observe that they have dedicated pages for specific industries—such as healthcare, finance, or education—you can create similar industry-focused pages on your site. For example, if your competitor has a comprehensive page targeting the healthcare sector with tailored solutions and testimonials, you can develop a dedicated healthcare solutions page that addresses the unique needs and challenges of that industry.
- **Location-Specific Pages:** If you see that your competitors have pages targeting specific geographic locations, especially those where your target audience resides, this can be a valuable insight. For instance, if they have separate pages for major cities or regions like New York, Los Angeles, or the Midwest, you could create localized content targeting these areas. This approach not only improves your local SEO but also

ensures that your content is highly relevant to the specific needs and preferences of customers in those locations.

User Experience Improvements:

- **Website Structure and Messaging:** By examining how a competitor's website is structured and the clarity of their messaging, you can gain insights into improving your own site. Notice how they present what they do, what they offer, and why potential customers should choose them over others. Is it immediately clear what services they provide? Do they effectively highlight their unique selling points? Apply these observations to make your own website easier to navigate and more compelling for visitors.
- **Ease of Understanding:** Ensure your site clearly communicates your offerings and value propositions. If competitors use clear, concise headings, bullet points, and engaging visuals to convey information, you can adopt similar strategies. Make it simple for potential customers to understand what you do and why you're the best choice.
- **Doing It Better:** Identify what your competitors do well and think about how you can do it even better. If their website excels in user experience, consider how you can enhance your own site's layout, design, and functionality to surpass theirs. This might include adding interactive elements, improving load times, or offering superior customer support options.

Engaging Content Types:

- **Content Marketing and SEO:** By observing a competitor's successful content strategy, such as regularly updated blogs, interactive infographics, or engaging video tutorials, you can identify which content types resonate most with your shared audience.
  1. **Blog Topics:** Take note of the topics they cover. Are there gaps or areas where you can provide more in-depth information?

2. **Posting Frequency:** Observe how often they post new content. Can you increase your frequency to keep your audience more engaged?
3. **Content Quality:** Evaluate the quality of their content. Can you create higher-quality visuals, more detailed articles, or more engaging videos?
4. **Improving Upon Their Strategies:** Think about how you can improve upon what your competitors are doing. Incorporate these insights into your own strategy to increase engagement and enhance your SEO efforts.

Effective Use of Social Proof:

- If competitors prominently feature customer testimonials, case studies, and user-generated content, it highlights the importance of social proof in your industry. You can benefit from incorporating similar elements into your website to build trust and credibility with potential customers.

Promotional Tactics:

- **Analyzing competitors' promotional strategies,** such as limited-time offers, loyalty programs, or bundled services, can provide insights into what appeals to your audience. By adopting and tailoring these tactics, you can enhance your own promotional efforts to better attract and retain customers.
- **Valuing Your Product or Service:** A lot of business and marketing companies push offers or sales. Do they help bring in business? Yes. However, you do have to think about the value of your product or service. Is the promotion potentially hurting the value your potential customer perceives your service/product to have?

Innovative Product Features:

- If a competitor is highlighting unique product features or services that address specific customer pain points, it indicates a market demand. You can leverage this information to innovate or emphasize similar features in your own offerings, ensuring you meet your audience's needs more effectively.

By visiting competitors' sites and understanding their strategies, you can gain valuable insights into what works and what doesn't for your target audience. Applying these learnings to your own website and marketing efforts can significantly enhance your competitive edge and drive your business success.

Chapter 4

EDUCATING VS SELLING TO CUSTOMERS

Imagine walking into a store where the salesperson bombards you with product details and prices the moment you step inside. It's overwhelming and off-putting, right?

Now, imagine a different scenario where the salesperson first takes the time to understand your needs, educates you about the options, and then guides you to the perfect product. This balanced approach not only makes the shopping experience pleasant but also increases the likelihood of making a sale.

One of the secrets to converting website visitors into loyal customers is balancing education and sales. Many businesses fall into the trap of overloading their customers with information in the wrong areas, leading to confusion and missed opportunities.

Digital marketing success isn't just about pushing your products or services; it's about building trust, providing value, and guiding your customers through their decision-making process. Effective content educates your audience while strategically positioning your products and services for sales. Whether you're a seasoned business owner or just starting out, mastering this balance is key to fostering trust and driving business growth.

Let's dive into the art of balancing education and sales in your digital marketing strategy.

Selling to Customers: The Role of Service and Product Pages

When it comes to your service or product pages, the goal is to sell. These pages should clearly communicate why customers should choose your business and what you offer that can solve their problems or fulfill their needs. Here's how to create compelling service and product pages:

- **Highlight Your Value Proposition:** Clearly state what sets your business apart from the competition. Focus on the benefits and unique features of your products or services.
- **Customer-Centric Content:** Write content that speaks directly to your customers' needs and desires. Use persuasive language that highlights how your offerings can solve their problems.
- **Clear and Compelling CTAs:** Include strong call-to-actions that guide visitors towards making a purchase or contacting you. Make sure these CTAs are visible and enticing.
- **Use Testimonials and Case Studies:** Include customer testimonials and case studies to build trust and credibility. Showcasing real-life success stories can be very persuasive.
- **Professional Design**: Ensure your service and product pages are well-designed, easy to navigate, and mobile-friendly. A professional look and feel can significantly impact customer perceptions.

Educating Customers: The Role of Your Blog

Your blog is the perfect platform for educating your customers. It's where you can demonstrate your expertise, provide valuable insights, and build trust with your audience. Here's how to do it effectively:

- **Identify Customer Pain Points**: Write about topics that address the challenges and questions your customers have. Use your knowledge and experience to provide solutions and guidance.

- **Create Valuable Content**: Offer detailed guides, how-tos, and informative articles. Ensure the content is well-researched and provides real value.
- **SEO Optimization**: Optimize your blog posts with relevant keywords to attract organic traffic. Use clear headings, internal links, and meta descriptions.
- **Engage Your Audience**: Encourage comments, questions, and discussions on your blog. Respond to comments to build a community around your content.

By focusing your educational efforts on your blog, you can establish your authority in your industry while providing your audience with the information they need to make informed decisions.

Balancing Education and Sales

Remember, the key is balance. Use your blog to educate and build trust, while your service and product pages should focus on selling and highlighting your value proposition. This approach ensures that you're providing the right information at the right time, helping to guide your customers through their journey from awareness to decision-making.

By clearly distinguishing between educational content and sales content, you can create a more effective and customer-centric website that not only attracts visitors but also converts them into loyal customers.

Example: Educating vs. Selling

To illustrate the importance of balancing education and sales, let's look at an example of a roofing company focusing on new roof installations. We will compare the incorrect approach and the correct approach to handling service/product pages and blog posts.

Example: Incorrect Approach

Scenario: A roofing company wants to sell roof installation services

Service Page Example

- **Page Title:** New Roof Installation
- **Section Headline:** What is a Roof Installation?
- **Page Body:** Roof installation involves removing the old roofing material and installing new shingles or other roofing products. The process includes laying down underlayment, ensuring proper ventilation, and securing the roofing materials in place. Proper installation is crucial for the longevity and effectiveness of your roof.

Types of Roofing Materials

There are various types of roofing materials available, including asphalt shingles, metal roofing, tile, and slate. Each material has its own benefits and drawbacks, and choosing the right one depends on your budget, climate, and aesthetic preferences.

- **Call to Action:**
  - Want to learn more about roof installations? Contact Us

Blog Post Example

- **Article Title:** The Benefits of Our New Roof Installation Services
- **Article Body:** Our new roof installation services ensure a durable and long-lasting roof for your home. Here are some benefits:

- High-Quality Materials: We use only the best materials to ensure your roof's durability.
  - Professional Installation: Our team of experts guarantees a perfect installation every time.
  - Comprehensive Warranty: We offer a warranty that covers both materials and labor.
- **Call to Action:**
  - Ready for a new roof? Contact Us

In this incorrect approach, the product page includes educational content that is better suited for a blog, and the blog post focuses on selling the product, which should be the focus of the product page. This can confuse visitors and dilute the effectiveness of the content.

Example: Correct Approach

Scenario: A roofing company wants to sell roof installation services

Service Page Example

- **Page Title:** New Roof Installation
- **Section Headline:** Why Choose Our Roof Installation Services?
- **Page Body:**
  - **High-Quality Materials:** We use only the best materials to ensure your roof's durability and longevity.
  - **Expert Installation:** Our experienced team guarantees a flawless installation that protects your home for years to come.
  - **Comprehensive Warranty:** We offer a warranty that covers both materials and labor, providing you with peace of mind.
  - **What Our Customers Say**
    - "We had our roof replaced by this company, and we couldn't be happier. The quality of work and professionalism were outstanding." - Sarah P.

- **Call to Action:**
  - Ready for a New Roof? Contact Us

Blog Post Example

- **Article Title:** How to Choose the Right Roofing Material for Your Home
- **Article Body:**
  - Choosing the right roofing material is essential for both the longevity of your roof and the overall appearance of your home. Here's a breakdown of the most popular materials:
  - Asphalt Shingles: Affordable and versatile, asphalt shingles are the most common roofing material in the U.S.
  - **Factors to Consider**
    - When selecting a roofing material, consider factors such as your budget, the climate in your area, and your home's architectural style.
- **Call to Action:**
  - Contact Us

In the correct approach, the product/service page focuses on selling by highlighting the unique features, benefits, and customer testimonials, with a clear call-to-action to buy.

The blog post educates the reader about choosing the right roofing material, providing valuable insights and tips without directly pushing a sale. This clear distinction helps guide visitors through their journey from learning about roofing options to making a purchase decision.

Developing Successful Promotional Strategies

Promotions are a powerful way to attract and retain customers. However, they need to be well-planned and targeted to be effective.

Understanding Different Types of Promotions

Different promotions can serve various purposes, from attracting new customers to retaining existing ones.

- **Discounts and Coupons**: Offer immediate savings to incentivize purchases.
- **Bundle Deals**: Encourage customers to buy more by offering a discount on bundled products.
- **Loyalty Programs**: Reward repeat customers to increase retention.
- **Limited-Time Offers**: Create urgency and encourage quick decisions.
- **Free Trials**: Allow potential customers to experience your product or service without commitment.

Evaluating the Impact of Promotions on Brand Perception

While promotions can drive sales, they can also impact how your brand is perceived.

Positive Impacts

- **Increased Sales**: Short-term boosts in revenue.
- **Customer Acquisition**: Attract new customers with enticing offers.
- **Customer Retention**: Keep existing customers engaged and loyal.

Negative Impacts

- **Perceived Value**: Frequent promotions can lead to customers expecting discounts, potentially undermining the perceived value of your products or services.
- **Brand Image**: Excessive discounts may damage the brand's premium image.

Customizing Promotions for Your Audience

Promotions should be tailored to meet the specific needs and preferences of your target audience.

Segmenting Your Audience

- Use data and insights to segment your audience based on demographics, behavior, and preferences.
- Develop targeted promotions for each segment to maximize engagement and conversion.

Testing and Iterating Promotional Tactics

Promotional strategies should be continuously tested and refined based on performance.

A/B Testing

- Test different promotional tactics to see which ones resonate most with your audience.
- Use the results to refine and optimize your strategies.

Performance Analysis

- Track key metrics such as conversion rates, average order value, and customer acquisition cost.

- Adjust your promotional tactics based on the data to improve effectiveness.

Case Studies of Successful Promotions

Case Study 1: E-commerce Store

Scenario: An online clothing retailer runs a seasonal sale offering 20% off all items for a limited time.

- **Results**: Significant increase in sales during the promotion period, with a boost in new customer acquisition and repeat purchases.

Case Study 2: SaaS Company

Scenario: A software company offers a 14-day free trial for their new productivity tool.

- **Results**: High conversion rate from trial users to paying customers, with increased customer retention due to the product's value demonstrated during the trial.

Case Study 3: Local Restaurant

Scenario: A restaurant launches a loyalty program offering a free meal after ten purchases.

- **Results:** Increased customer visits and higher average spend per visit, with strong customer loyalty and positive word-of-mouth referrals.

Balancing education and sales, along with developing effective promotional strategies, is key to converting visitors into loyal customers.

By understanding the roles of different content types and tailoring promotions to your audience, you can create a comprehensive marketing strategy that drives growth and profitability.

Chapter 5

UNDERSTANDING SEO

Ever wondered why some websites always seem to pop up at the top of your search results while others are buried pages deep? It's no accident—those top sites have mastered the art of SEO.

Search Engine Optimization (SEO) is like having a treasure map that leads potential customers straight to your website. It involves optimizing your site so search engines (Google) can find, and understand the site and rank it higher in search results. Higher rankings mean more visibility, more traffic, and more opportunities for conversions, leads, or sales.

Digital marketing isn't just about having a beautiful website or running compelling ads; it's about making sure your target audience can find you when they need you most. We will be diving into key concepts such as keyword research, on-page SEO, off-page SEO, and technical SEO.

Ready to unlock the secrets of SEO and propel your business to the top of search results? Let's dive into the fundamentals of search engine optimization.

The User Comes First

Your SEO research, strategy, and implementation should always be about the user first and Google second. That's why we discussed "Understanding Your Ideal Customer" way back in Chapter 2. This is where it really comes together.

By understanding who your customers are, what they need, and how they search for information, you can create content that meets their needs and naturally aligns with SEO best practices.

This user-first approach not only improves your rankings but also enhances the user experience, leading to higher engagement and conversion rates.

Search Engine Algorithms

To grasp SEO, you first need to understand search engine algorithms. These complex systems determine which pages appear in search results and in what order. Google, the dominant search engine, frequently updates its algorithm to improve user experience and deliver the most relevant results.

- **Major Algorithm Updates**
- **Panda:** Focuses on content quality, penalizing sites with thin or duplicate content.
- **Penguin:** Targets sites with manipulative link practices.
- **Hummingbird:** Enhances understanding of search queries' intent.
- **RankBrain:** Uses machine learning to improve search results relevance.

The Evolution of Google's Algorithm

Over the years, Google's algorithm has evolved to prioritize user experience. Key updates have shifted the focus towards high-quality content, mobile-friendliness, and secure websites. Staying updated with these changes will better prepare for the possibility that you'll have to pivot your SEO strategy to continue to maintain and improve your site's SEO performance.

Google Indexing Process

For your site to appear in search results, Google needs to discover, crawl, and index it.

- **Discovery**
  - Google uses bots, also known as spiders, to find new and updated pages by following links.
- **Crawling**
  - During crawling, Googlebot visits your pages, reads the content, and analyzes the structure.
- **Indexing**
  - In this phase, the content is processed and stored in Google's database. Factors like keywords, freshness, and relevance are considered.

The Learning Phase

Google often places new sites and pages in a "learning phase" where it evaluates their relevance and quality. During this time, your SEO performance may fluctuate before you see relatively steady ranking patterns.

Factors Influencing SEO Performance

SEO involves both controllable and uncontrollable elements.

Uncontrollable Elements

- **Algorithm Changes:** Google's updates can impact your rankings.
- **Indexing Speed:** Varies based on Google's schedule.

- **Competition:** High competition for keywords can affect visibility.

Controllable Elements

- **Content Quality:** Regularly update with high-quality, relevant content.
- **Technical SEO:** Ensure your site is technically sound.
- **User Experience:** Optimize for mobile, improve site speed, and ensure easy navigation.
- **Backlinks:** Build a strong backlink profile from reputable sites.

Keyword Research and Analysis

- **Brainstorm Seed Keywords:** Think about broad topics related to your business. Use tools like Google Keyword Planner to expand these into a list of potential keywords.
- **Analyze Keyword Metrics:** Look at search volume, keyword difficulty, and competition using tools like Ahrefs and SEMrush.
- **Identify Long-Tail Keywords:** Focus on specific, intent-driven phrases like "best running shoes for flat feet."
- **Check Competitor Keywords:** See what keywords your competitors are ranking for using SEMrush.
- **Organize and Prioritize Keywords:** Create a keyword strategy that aligns with your business goals, prioritizing based on relevance, search volume, and competition.

On-Page SEO Best Practices

- **Title Tags and Meta Descriptions:** Include primary keywords and keep them concise and compelling.
- **Header Tags (H1, H2, H3):** Use them to structure your content and include keywords naturally.

- **Content Quality:** Ensure it is informative and engaging, using keywords naturally without overstuffing.
- **Internal Linking:** Link to relevant pages within your site to help with navigation and SEO.
- **URL Structure:** Use clean, descriptive URLs, including keywords if possible.
- **Image Optimization:** Use descriptive file names and alt text, compress images for faster loading.

Off-Page SEO Strategies

- **Building Backlinks:** Guest post on reputable sites, engage in industry forums, and comment on blogs.
  - Many marketing companies do not invest heavily in backlink building, or they do it minimally. It's a time-consuming process, and if you're just starting out or your business is not well-known, securing quality links beyond foundational ones like Yelp, YP, or your local chamber of commerce can be a challenge. Don't get me wrong—if you can secure quality backlinks, it certainly helps in the broad spectrum of SEO. However, you don't need a ton to rank in Google. There are more ways to rank your site than just backlinks.
  - **Do not waste your hard-earned money on buying links.** It's tempting, I know, but trust me, they are usually all garbage and will do more harm than good or nothing at all. Spend your money on solid SEO strategies that you know will benefit your business.
- **Social Media Engagement:** Share your content on social platforms and engage with your audience regularly.
- **Local Citations:** Ensure your business information is consistent across directories and get listed on local business directories.
- **Content Marketing:** Create high-quality, shareable content and collaborate with influencers and bloggers.

Real-world Examples

Example 1: Local Business A local bakery optimizes for "gluten-free bakery [city name]" and sees a spike in traffic within days due to high local demand and low competition.

Example 2: E-commerce Site An online store selling handmade crafts starts blogging about DIY crafts. It takes about three months to see significant traffic increase due to competitive keywords but gains a loyal following and higher sales over time.

Example 3: Service-based Business A home cleaning service targets long-tail keywords like "eco-friendly home cleaning service [city name]" and begins to rank within a month due to the specificity and lower competition of the keywords. Over six months, this strategy leads to a consistent increase in inquiries and bookings.

Setting Realistic Expectations

SEO isn't just about ranking high in the Search Engine Results Pages (SERPs). It's about attracting the right traffic—visitors who are genuinely interested in what you offer. By focusing on long-tail keywords and high-quality content, you can target users with clear intent, increasing the likelihood of conversions.

Implementing a comprehensive SEO strategy is a continuous process of optimization and adaptation. It involves not only understanding how search engines rank sites but also keeping up with algorithm changes and evolving best practices. With dedication and the right strategies, you can significantly improve your search engine rankings and drive more targeted traffic to your site.

Ready to become an SEO master? Let's move on to discuss all the elements of SEO in detail.

Chapter 6

KEYWORD RESEARCH AND ANALYSIS

Think of setting off on a treasure hunt without a map or clues. The chances of finding the treasure would be slim to none. This is what happens when you embark on an SEO campaign without proper keyword research and analysis. Keywords are the compass that guides your content strategy, helping you understand what your audience is searching for and how to attract them to your site.

Keyword research is the bedrock of SEO, providing insights into the terms and phrases your potential customers use when searching for products or services like yours. This chapter will guide you through the steps of brainstorming seed keywords, analyzing keyword metrics, identifying long-tail keywords, and understanding the competitive landscape.

Digital marketing is not just about creating content; it's about creating the right content that resonates with your audience and meets their search intent. We will cover various tools and techniques for keyword research, even exploring strategies for when you don't have access to advanced keyword research tools.

By the end of this chapter, you'll be equipped with the knowledge and tools to develop a robust keyword strategy that aligns with your business goals and drives relevant traffic to your site. Whether you're a seasoned business owner or just starting out, mastering keyword research and analysis is essential for optimizing your content and maximizing your SEO efforts.

Ready to discover the keywords that will unlock your business's potential and lead you to digital marketing success? Let's dive into the world of keyword research and analysis.

The Difference Between Keywords and Queries

Before diving deeper, it's important to understand the distinction between keywords and queries:

- **Keyword:** A specific word or phrase that digital marketers and content creators use to optimize content for search engines. Keywords are chosen based on what users might type into a search engine to find relevant information, products, or services.
  - **Example:** "best running shoes" is a keyword that a sports retailer might target in their content to attract users searching for running shoes.
- **Query:** The actual word or phrase that a user types into a search engine. Queries are how users express their need for information. They can be simple or complex and often reflect the intent behind the search.
  - **Example:** "What are the best running shoes for marathon training?" is a query entered by a user into a search engine.

Understanding this difference helps you align your keyword strategy with the actual language and intent of your target audience.

The Importance of Keyword Research

Keyword research helps you understand what your potential customers are searching for and how to attract them to your site. It involves finding the terms and phrases that people use in search engines related to your products, services, or content. This process allows you to:

- **Identify Market Demand:** Understand what topics are popular and relevant to your audience.

- **Target Relevant Traffic:** Attract visitors who are interested in what you offer.
- **Increase Conversions:** Reach users who are more likely to convert into customers.

Understanding Key Concepts in Keyword Research

- **Search Volume:** Search volume refers to the number of times a particular keyword is searched for in a given period, usually monthly. High search volume indicates a popular keyword, but it also often means higher competition. However, not all valuable keywords have high search volumes.
- **Buyer Intent:** Buyer intent (or search intent) is the purpose behind a search query. It indicates what the user is looking to achieve. Understanding buyer intent is crucial because it helps you target keywords that are more likely to convert. There are different types of intent:
  - **Transactional Intent:** The user intends to make a purchase. Keywords include "buy," "discount," "deal," "cheap," and specific product names.
    - **Example:** "buy eco-friendly water bottles," "discount running shoes"
  - **Commercial Intent:** The user is researching products or services with the intent to purchase soon. Keywords include "best," "top," "review," and "comparison."
    - **Example:** "best running shoes for flat feet," "top digital marketing agencies"
  - **Informational Intent:** The user is looking for information or answers to questions. Keywords include "how," "what," "why," and "guide."
    - **Example:** "how to choose running shoes," "what are eco-friendly products"
- **Why Intent Matters:** Knowing the intent behind keywords helps you create content that meets the user's needs and drives conversions. For instance, targeting transactional

keywords can directly lead to sales, while informational keywords can attract visitors early in their buying journey and build trust.

Relevance and Intent: The Cornerstones of Effective SEO Strategies

Relevance and intent are critical factors in keyword research and SEO. Understanding and prioritizing these elements can significantly impact the success of your SEO strategy.

- **Relevance:** Relevance refers to how closely a keyword matches the content and purpose of your website. Using relevant keywords ensures that the visitors you attract are genuinely interested in what you offer.
  - **Example:**
    - If you own a bakery, targeting the keyword "best bakery in [city]" is more relevant than "baking recipes," which may attract visitors looking for DIY baking tips rather than those wanting to purchase baked goods.
- **Intent:** Intent reveals the user's purpose behind a search query. Different types of intent (informational, commercial, transactional) require different approaches in content creation and keyword targeting.
  - **Example:**
    - A user searching for "how to bake sourdough bread" has informational intent and is likely seeking a recipe or guide. In contrast, someone searching for "buy sourdough bread near me" has transactional intent and is ready to make a purchase.

Integrating Relevance and Intent into Your Keyword Strategy

- **Identifying High-Intent Keywords:** Focus on keywords with clear buyer intent to attract users who are ready to convert. These keywords often include terms like "buy," "discount," "best," and specific product names.
  - **Example:**
    - **High-intent keyword:** "buy eco-friendly water bottles"
    - **Low-intent keyword:** "what are eco-friendly products"
- **Balancing High and Low Search Volume Keywords:** While high search volume keywords can drive a lot of traffic, low search volume keywords can be just as valuable if they have high buyer intent. Don't overlook keywords with low search volume but high relevance and intent.
  - **Example:**
    - **High search volume, low intent:** "running shoes"
    - **Low search volume, high intent:** "best running shoes for flat feet"
- **Creating Content that Matches User Intent:** Develop content that aligns with the different types of user intent. This ensures you meet the needs of users at various stages of their buying journey.
  - **Example:**
    - **Informational content:** Blog post titled "How to Choose the Right Running Shoes"
    - **Commercial content:** Product comparison page titled "Top 10 Running Shoes"
    - **Transactional content:** Product page titled "Buy the Best Running Shoes for Flat Feet"

Practical Examples for Different Scenarios

Local Business Example

Business: Local Bakery **Goal:** Attract local customers looking for fresh bread and pastries.

- **Brainstorm Seed Keywords:**
  - Fresh bread, bakery, pastries, local bakery, bakery near me
- **Expand and Analyze Keywords:**
  - Use Google Autocomplete and Related Searches to find: "best bakery in [city]," "fresh bread near me," "local pastries," "artisan bakery [city]"
- **Identify Long-Tail Keywords:**
  - "Best bakery for sourdough bread in [city]"
- **Check Competitor Keywords:**
  - Visit local competitor websites and note keywords used in their titles, descriptions and throughout their site.
- **Organize and Prioritize Keywords:**
  - **High priority:** "best bakery in [city]"
  - **Medium priority:** "fresh bread near me"
  - **Low priority:** "local pastries"
- **Consider Intent and Relevance:**
  - **Transactional intent:** "buy fresh bread [city]"
  - **Informational intent:** "how to make sourdough bread"

Local Business - Creating Content Based on Intent

- **Informational Content:**
  - **Blog Post:** "How to Choose the Best Bakery in [City]"
    - **Content:** Provide tips on what to look for in a bakery, including the quality of ingredients, variety of products, and customer reviews.
  - **Blog Post:** "The Benefits of Eating Gluten-Free Pastries"
    - **Content:** Explain the health benefits of gluten-free pastries, with a spotlight on your bakery's gluten-free options.
- **Commercial Content:**
  - **Service Page:** "Best Bakery in [City]"

- **Content:** Highlight your bakery's unique selling points, such as daily fresh bread, artisan techniques, and customer testimonials.
  - **Service Page:** "Gluten-Free Bakery Near Me"
    - **Content:** Emphasize your gluten-free product range, quality ingredients, and customer satisfaction. Include a CTA like "Visit Us Today."
- **Transactional Content:**
  - **Landing Page:** "Buy Fresh Bread in [City]"
    - **Content:** Use local SEO, include information about your bread-making process, and customer reviews. Strong CTAs like "Order Now" or "Visit Our Bakery."
  - **Landing Page:** "Order Pastries Near Me"
    - **Content:** Provide details on your pastry selection, special offers, and easy ordering options. Include CTAs such as "Order Online" or "Call to Place Your Order."

E-Commerce Business Example

Business: Online Eco-Friendly Product Store **Goal:** Attract customers looking for sustainable and eco-friendly products.

- **Brainstorm Seed Keywords:**
  - Eco-friendly products, sustainable products, green products, reusable products
- **Expand and Analyze Keywords:**
  - Use Google Autocomplete and Related Searches to find: "best eco-friendly products," "sustainable home goods," "reusable water bottles"
- **Identify Long-Tail Keywords:**
  - "Best eco-friendly cleaning products for home"
- **Check Competitor Keywords:**
  - Analyze competitor sites for keywords like "zero waste products," "biodegradable products"
- **Organize and Prioritize Keywords:**
  - **High priority:** "best eco-friendly products"
  - **Medium priority:** "sustainable home goods"

- o **Low priority:** "reusable water bottles"
- **Consider Intent and Relevance:**
  - o **Transactional intent:** "buy reusable water bottles"
  - o **Informational intent:** "benefits of eco-friendly products"

E-Commerce Business - Creating Content Based on Intent

- **Informational Content:**
  - o **Blog Post:** "The Ultimate Guide to Eco-Friendly Products for Your Home"
    - **Content:** Provide comprehensive information on eco-friendly products, including benefits, types, and how to choose the best ones for your home.
  - o **Blog Post:** "How to Transition to a Zero-Waste Lifestyle"
    - **Content:** Offer tips and advice on starting a zero-waste lifestyle, highlighting products that can help in the transition.
- **Commercial Content:**
  - o **Product Page:** "Best Eco-Friendly Cleaning Products for Home"
    - **Content:** Detail the features and benefits of your cleaning products, customer reviews, and eco-friendly certifications.
  - o **Product Page:** "Affordable Zero Waste Products"
    - **Content:** Highlight your affordable product range, benefits of zero-waste living, and special offers. Include a CTA like "Shop Now."
- **Transactional Content:**
  - o **Landing Page:** "Buy Eco-Friendly Cleaning Products"
    - **Content:** Emphasize the quality, effectiveness, and environmental benefits of your products. Include strong CTAs such as "Buy Now" or "Add to Cart."
  - o **Landing Page:** "Order Sustainable Home Goods"
    - **Content:** Provide detailed product descriptions, benefits, and customer testimonials. Include CTAs like "Order Today" or "Free Shipping on Orders Over $50."

Service-Based Business Example

Business: Accounting Firm **Goal:** Attract clients looking for accounting services.

- **Brainstorm Seed Keywords:**
  - Accounting services, tax preparation, bookkeeping, CPA, financial planning
- **Expand and Analyze Keywords:**
  - Use Google Autocomplete and Related Searches to find: "best accounting services," "tax preparation near me," "bookkeeping for small businesses," "certified public accountant"
- **Identify Long-Tail Keywords:**
  - "Best accounting services for small businesses"
  - "Affordable tax preparation services"
- **Check Competitor Keywords:**
  - Analyze competitor sites for keywords like "top CPA firms," "tax advisors," "professional bookkeeping services"
- **Organize and Prioritize Keywords:**
  - **High priority:** "tax preparation services near me"
  - **Medium priority:** "accounting services for small businesses"
  - **Low priority:** "financial planning advice"
- **Consider Intent and Relevance:**
  - **Transactional intent:** "hire an accountant for tax preparation"
  - **Informational intent:** "how to choose an accountant for small business"

Service-Based Business - Creating Content Based on Intent

- **Informational Content:**
  - **Blog Post:** "How to Choose the Right Accountant for Your Small Business"

- **Content:** Provide tips on what to look for in an accountant, questions to ask during consultations, and the benefits of hiring a professional.
  - o **Blog Post:** "Top Benefits of Professional Bookkeeping Services"
    - **Content:** Explain the advantages of professional bookkeeping, such as accuracy, time savings, and compliance with tax laws.
- **Commercial Content:**
  - o **Service Page:** "Best Accounting Services for Small Businesses"
    - **Content:** Detail the specific services offered for small businesses, including tax preparation, bookkeeping, and financial planning. Highlight case studies or testimonials from small business clients.
  - o **Service Page:** "Affordable Tax Preparation Services"
    - **Content:** Emphasize the affordability and quality of your tax preparation services. Include a clear call-to-action, such as "Schedule Your Consultation Today."
- **Transactional Content:**
  - o **Landing Page:** "Tax Preparation Services Near Me"
    - **Content:** Focus on local SEO by including location-specific keywords. Highlight your expertise, certifications, and the benefits of choosing your firm for tax preparation. Include strong calls-to-action like "Book Now" or "Get a Free Quote."
  - o **Landing Page:** "Hire an Accountant for Tax Preparation"
    - **Content:** Provide detailed information on the tax preparation process, the experience of your team, and the benefits of professional tax services. Include testimonials and a CTA to schedule a consultation.

Effective keyword research is a game changer for any successful SEO strategy. By identifying the right keywords, considering their relevance and intent, and creating content that meets user needs, you can attract relevant traffic, increase conversions, and grow your business.

Whether you have access to advanced keyword research tools or not, these strategies and examples will help you uncover valuable keywords and create a strong foundation for your SEO efforts.

Now let's explore Technical SEO Best Practices to help you optimize your content and boost your search engine rankings.

Chapter 7

TECHNICAL SEO: OPTIMIZING YOUR WEBSITE

Imagine building a magnificent skyscraper with all the right aesthetics, only to find that its foundation is shaky and unstable. This is what happens when a website lacks proper technical SEO— no matter how great the content or how well you implement on-page strategies, your site won't reach its full potential without a solid technical foundation.

Let's dive into the foundational aspects of technical SEO that ensure your website is both crawlable and indexable by search engines. Technical SEO is the backbone that supports all other SEO efforts. It involves optimizing the infrastructure of your website to improve its performance, ensuring that search engines can easily find, crawl, and index your content.

Digital marketing isn't just about flashy designs and compelling content; it's about ensuring that every element of your website works together seamlessly to provide the best possible experience for both users and search engines.

Ready to fortify the foundation of your website and unlock its true potential? Let's dive into the world of technical SEO.

The Importance of Technical SEO

Technical SEO ensures that your website is built in a way that search engines can efficiently crawl and index your content. Without it, even the best content might not rank well. Technical SEO focuses on improving the backend elements of your site, ensuring that it meets the technical requirements of search engines and provides a seamless user experience.

Key Components of Technical SEO

- Website Speed and Performance
- Mobile-Friendliness
- Crawlability
- Indexability
- XML Sitemaps
- Structured Data
- HTTPS
- Canonicalization
- URL Structure
- Robots.txt File
- 404 Pages and Redirects
- International SEO

1. Website Speed and Performance

Why It Matters

Website speed is incredibly important for both user experience and search engine rankings. Think about it—how often do you wait for a slow website to load? Probably not very often. A fast website keeps your visitors happy, reduces bounce rates, and boosts engagement and conversions.

Google wants to serve the best and most relevant sites to its users, and fast-loading websites are a key part of that. In today's competitive market, nobody has the patience for a sluggish site when faster alternatives are just a click away.

Best Practices

- **Optimize Images:** Compress images using tools like TinyPNG or ImageOptim.
- **Enable Browser Caching:** Use caching plugins to store static files and reduce load times.
- **Minify CSS, JavaScript, and HTML:** Remove unnecessary characters from code to improve load speed.
- **Use a Content Delivery Network (CDN):** Distribute content across multiple servers to reduce latency.

2. Mobile-Friendliness

Why It Matters

With the majority of searches now happening on mobile devices, having a mobile-friendly website is essential. Google uses mobile-first indexing, meaning it predominantly uses the mobile version of the content for indexing and ranking. Even if your ideal customer typically searches for your services on a desktop, prioritizing mobile-friendliness is crucial because Google evaluates the mobile version of your site first, impacting your overall search engine rankings.

Best Practices

- **Responsive Design:** Ensure your website adapts to different screen sizes.
- **Touch-Friendly Elements:** Make sure buttons and links are easily tappable on mobile devices.
- **Optimize for Fast Loading:** Use AMP (Accelerated Mobile Pages) to speed up mobile pages.
- **Example of Optimizing for Mobile**

Consider the design of buttons on your website. On a desktop, a smaller button might work fine, but on a cell phone, it needs to be large enough to be easily tapped with a finger. Here's an example of the backend code for a button:

Before Optimization:

```
<button style="width:100px;">Click Me</button>
```

This button has a fixed width, which might be too small for mobile users to tap comfortably.

After Optimization:

```
<button style="width:100%; padding:15px;">Click Me</button>
```

In this optimized version, the button is set to take the full width of the screen with added padding, making it easier for mobile users to tap. This improves the user experience on mobile devices by ensuring buttons are large enough to be easily clickable.

Important Consideration

When determining your ideal customer, you might find that they primarily search for your type of services on a desktop. This could lead you to put less emphasis on the mobile version of your site. However, focus on mobile-friendliness because of Google's mobile-first indexing.

Ensuring your site is mobile-friendly not only caters to mobile users but also helps improve your search engine rankings, ultimately attracting more traffic and potential customers.

3. Crawlability

Crawlability refers to the ability of search engines to access and navigate through your website's pages. When search engines like

Google crawl your site, they follow links to discover new content and update their indexes.

If search engines can't crawl your site, they can't index it, which means your site won't appear in search results. Ensuring that your website is crawlable is top priority for SEO because it allows search engines to properly index your content and make it available to users in search results.

4. Indexability

Indexability is the ability of a search engine to analyze and add a page to its index which allows that page to rank. Without proper indexability, your pages won't appear in search results or if they do they might not appear properly.

- **Robots.txt:** Ensure your robots.txt file doesn't block important pages.
- **Meta Tags:** Use meta tags to control indexing (e.g., noindex for pages you don't want indexed).
- **Canonical Tags:** Prevent duplicate content issues by using canonical tags.

5. XML Sitemaps

XML sitemaps help search engines understand the structure of your website and find all your pages. It's a roadmap for search engines to crawl your site more effectively.

Create an XML Sitemap

For WordPress Websites: Using SEO plugins simplifies the process of creating a sitemap. Usually, these SEO plugins automatically generate a sitemap for you, ensuring all your pages are included.

Step-by-Step Example:

- **Install an SEO plugin:** Go to the Plugins section in your WordPress dashboard, search for an SEO plugin, and install it.
- **Activate the Plugin:** Once installed, activate the SEO plugin.
- **Generate Sitemap:** The plugin typically generates a sitemap automatically. You can find it by navigating to the plugin's settings in your WordPress dashboard. A lot of sitemaps you can find by typing your domain and then the sitemap such as (https://www.yoursite.com/sitemap.xml)

For Non-WordPress Websites:

You can use various online tools or CMS features to generate an XML sitemap. Ensure that the generated sitemap includes all essential pages and is structured correctly.

Submit to Google Search Engines:

Submitting your sitemap to Google Search Console is crucial for efficient indexing.

Step-by-Step Guide:

- **Login to Google Search Console:** If you don't have an account, create one and verify your website.
- **Navigate to Sitemaps:** In the left-hand menu, click on "Sitemaps."
- **Enter Sitemap URL:** Input the URL of your sitemap (e.g., https://www.example.com/sitemap.xml) and click "Submit."

Keep It Updated

Regularly updating your sitemap ensures search engines have the most current view of your site's structure. This is particularly important when you add or remove pages.

Automated Updates for WordPress:

If you use an SEO plugin, the plugin typically updates the sitemap automatically whenever you make changes to your site.

Why This Matters:

Submitting your sitemap helps Google discover new and updated content faster, improving your site's crawlability and indexability but let's go deep into this.

- **Improved Crawlability:** A sitemap helps search engine crawlers understand the structure of your website and discover all the pages, especially those that might not be easily found through internal linking. For instance, if you have orphan pages (pages not linked from anywhere on your site), a sitemap ensures they are still discovered.
- **Faster Indexing:** With a sitemap, new content and updates on your site can be discovered and indexed more quickly by search engines. For example, if you run a news website, having a regularly updated sitemap ensures your latest articles are indexed promptly.
- **Enhanced Visibility for Non-Linked Pages:** If you have pages that are not linked internally or are buried deep within your site's structure, a sitemap ensures these pages are still accessible to search engines. Consider a large e-commerce site with deep product categories; a sitemap ensures all products are indexed.
- **Better Handling of Large Sites:** For websites with a complex structure or a large number of pages, a sitemap helps search engines navigate the site more efficiently, ensuring that all relevant content is considered for ranking. Think of a site like

Amazon with millions of product pages; a sitemap is crucial for ensuring all products are discoverable.

- **Inclusion of Metadata:** Sitemaps can include metadata about pages, such as the last update time, change frequency, and importance relative to other URLs on the site. This information helps search engines understand the significance and relevance of each page. For example, if you have a seasonal product page, indicating its change frequency can help search engines prioritize it during relevant times.
- **Indication of Canonical URLs:** In cases where you have duplicate content, a sitemap can indicate the preferred version of a URL, helping to consolidate link equity and avoid duplicate content issues. For example, if you have multiple pages with similar content, specifying the canonical version helps search engines understand which one to rank.
- **Informing Search Engines of Multimedia Content:** Sitemaps can also include information about video, images, and news content, helping search engines understand and index these media types properly. For a site with a lot of video content, an XML sitemap with video metadata ensures that all videos are indexed and appear in search results.

6. Structured Data (Schema Markup)

Structured data helps search engines better understand the content of your pages. It can enhance your listings with rich snippets, increasing your click-through rate (CTR). This important aspect is often missed even by seasoned marketers because it can be complex to create. However, with AI tools, you can easily generate the necessary code. Be careful to test the code, as incorrect information can harm your rankings.

Schema Markup

Schema markup is a type of structured data that uses a specific vocabulary to help search engines understand the context and meaning of your content. By adding schema markup to your HTML, you provide search engines with additional information that can improve the display of your search results.

Use schema markup relevant to your content:

There are many different schema markups you can implement depending on your business.

- Article: For news articles, blog posts, and similar content.
- Product: For e-commerce product pages.
- FAQ: For frequently asked questions and answers.

Rich Snippets

Rich snippets are the enhanced search results that include additional information extracted from your structured data. They make your listings more attractive and informative in the search results, potentially increasing your CTR.

Examples of rich snippets:

- Review snippets: Display star ratings and review counts.
- Product snippets: Show product price, availability, and other details.
- Recipe snippets: Include cooking time, ingredients, and calorie information.

Schema Markup Examples

Here are some examples of schema markup code. Keep in mind that some of these should be applied sitewide (on every page), while others are more appropriate for specific pages. Additionally, the actual schema markup used on your site may be more extensive than the examples provided.

Organization Schema

The organization schema provides information about a general organization or business entity.

```
<script type="application/ld+json">

{

"@context": "http://schema.org",

"@type": "Organization",

"name": "Example Company",

"url": "http://www.example.com",

"logo": "http://www.example.com/images/logo.png",

"contactPoint": {

"@type": "ContactPoint",

"telephone": "+1-800-555-1212",

"contactType": "Customer Service",

"areaServed": "US",

"availableLanguage": "English"

},

"sameAs": [

"http://www.facebook.com/example",

"http://www.twitter.com/example",
```

```
]

}
```

`</script>`

FAQ Schema

The FAQ schema helps search engines understand that the content is a list of questions and answers.

```
<script type="application/ld+json">

{

"@context": "http://schema.org",

"@type": "FAQPage",

"mainEntity": [{

"@type": "Question",

"name": "What is your return policy?",

"acceptedAnswer": {

"@type": "Answer",

"text": "You can return any item within 30 days of purchase."

}

}]

}
```

`</script>`

Local Business Schema

The local business schema provides detailed information about a local business, such as its physical location, opening hours, and geographical coordinates.

```
<script type="application/ld+json">

{

"@context": "http://schema.org",

"@type": "LocalBusiness",

"name": "Example Local Business",

"image": "http://www.example.com/images/logo.png",

"@id": "http://www.example.com",

"url": "http://www.example.com",

"telephone": "+1-800-555-1212",

"address": {

"@type": "PostalAddress",

"streetAddress": "123 Main St",

"addressLocality": "Anytown",

"addressRegion": "CA",

"postalCode": "90210",

"addressCountry": "US"

},

"geo": {

"@type": "GeoCoordinates",
```

"latitude": 00.00000,

"longitude": 000.0000

},

"openingHoursSpecification": {

"@type": "OpeningHoursSpecification",

"dayOfWeek": [

"Monday",

"Tuesday",

"Wednesday",

"Thursday",

"Friday"

],

"opens": "09:00",

"closes": "17:00"

},

"sameAs": [

"http://www.facebook.com/example",

"http://www.twitter.com/example",

]

}

</script>

Schema Markup and Generate Rich Snippets

- **Choose the Right Schema Markup:**
  - ○ Select the appropriate schema type for your content. Refer to schema.org for a comprehensive list of schema types and properties.
- **Generate the Code:**
  - ○ Use online tools or AI-based generators to create the necessary schema markup code. Ensure that the generated code accurately represents your content.
- **Add the Code to Your HTML:**
  - ○ Insert the schema markup into the relevant sections of your web pages. This can usually be done within the HTML header or within specific content blocks.
    - ▪ If you have a WordPress site you can use a plugin to inject the code. If you can't use a plugin then you can implement the schema through Google Tag Manager if you have that setup and connected to your site.
- **Validate Your Markup:**
  - ○ Use the schema.org validator or Google's Rich Results Test to check for errors and ensure your markup is correctly implemented.
- **Monitor and Adjust:**
  - ○ After implementation, monitor your search results and CTR. Adjust the markup as needed to improve performance.

The benefits of implementing schema markup extend beyond improved visibility. The structured data can lead to better user experiences, higher customer engagement, and increased trust in your brand. This can translate into more conversions and a higher return on investment (ROI) for your SEO efforts.

Investing the time and resources to properly implement and maintain schema markup is a strategic decision that can pay off significantly. Not only will it help you stay competitive but it will also position your business for long-term success by ensuring that your website is fully optimized for search engines and user-friendly.

Embracing schema markup is not just about technical SEO; it's about creating a more effective, engaging, and profitable online presence. Make the effort to implement it correctly, and you'll likely see substantial benefits for your business.

7. HTTPS

HTTPS is a security protocol that ensures data between your website and the user is encrypted. Google considers HTTPS as a ranking signal, making it important for SEO.

Example: Before Optimization:

<a href="http://www.example.com">Visit our site</a>

Example After Optimization:

<a href="https://www.example.com">Visit our site</a>

What Happens If You Don't Have an HTTPS Site?

- **Security Risks:** Without HTTPS, data transmitted between your website and users is not encrypted, making it vulnerable to interception and attacks. This can lead to data breaches, loss of sensitive information, and a lack of trust from your users.
- **Negative SEO Impact:** Google uses HTTPS as a ranking signal. Websites without HTTPS may rank lower in search engine results, reducing visibility and organic traffic. Additionally,

browsers like Chrome display a "Not Secure" warning for HTTP sites, which can deter visitors and affect your site's credibility.

  o This is one of the biggest reasons from a customer standpoint. No one wants their business site to show that it's potentially not trustworthy.
- **Compatibility Issues:** Many modern web features and APIs require HTTPS to function properly. Without it, you may experience compatibility issues, and some features may not work as intended.

By installing an SSL certificate, redirecting HTTP to HTTPS, and updating internal links, you can provide a safer, more reliable experience for your users and boost your site's performance in Google search results.

8. Canonicalization

Canonicalization helps prevent duplicate content issues by specifying the preferred version of a page. It tells search engines which version of a URL to index.

- **Use Canonical Tags:** Add canonical tags to pages with duplicate content.
- **Set Preferred Domain:** Choose a preferred domain (www or non-www) and stick to it.

Canonical Tag Example

Imagine you have an e-commerce site where the same product can be accessed through different URLs due to various filters or parameters.

- URL 1: http://www.example.com/product?color=red
- URL 2: http://www.example.com/product?size=large
- URL 3: http://www.example.com/product

Although these URLs point to the same product page, they have different parameters for color and size, leading to duplicate content issues.

Using a Canonical Tag

To avoid duplicate content problems and consolidate the ranking signals to a single preferred URL use the canonical tag.

Preferred URL: http://www.example.com/product

<link rel="canonical" href="http://www.example.com/product" />

9. URL Structure

A clean and descriptive URL structure makes it easier for search engines and users to understand the content of your pages.

- **Keep URLs Short and Descriptive:** Use concise URLs that describe the content.
- **Use Hyphens:** Separate words with hyphens rather than underscores.
- **Include Keywords:** Incorporate relevant keywords into URLs.

Example Before Optimization:

https://www.example.com/page123

Example After Optimization:

https://www.example.com/hvac-installation-services

Local Service URL Example:

http://www.example.com/locations/new-york/plumbing/

This is an example of a strategic URL structure because it includes relevant keywords like "locations," "New York," and "plumbing," which helps search engines understand the content and relevance of the page and including the location in the URL (e.g., "new-york") makes it clear to both search engines and users that this page is specifically about plumbing services in New York. This can improve local SEO and make the page more likely to appear in local search results.

By attracting users specifically looking for plumbing services in New York, the URL helps bring more qualified leads to the site. These visitors are more likely to convert into customers, boosting the business's ROI.

10. Robots.txt File

The robots.txt file instructs search engine crawlers which pages or sections of your site should not be crawled. It helps manage crawl budget and keep unwanted pages out of search engines.

- **Block Unimportant Pages:** Use robots.txt to block pages like admin sections.
- **Allow Essential Pages:** Ensure that important pages are not accidentally blocked.
- **Regularly Check:** Review and update your robots.txt file as needed.

This is the typical code you would see in a robots.txt file:

User-agent: *

Disallow: /admin/

Disallow: /login/

Allow: /

11. 404 Pages and Redirects

404 errors typically occur when a page on your site was deleted but there is a link or the page is shown in search engines and a user clicks that link or search result, they will be taken to a page that cannot be found a "404 page". Handling these correctly can improve user experience and preserve link equity.

Link equity, also known as "link juice," refers to the value or authority passed from one webpage to another through hyperlinks. This value is an important factor in search engine optimization (SEO) because it influences the ranking of a website's pages in search engine results.

- **Custom 404 Page:** Create a helpful 404 page that guides users back to your site.
- **Use 301 Redirects:** Redirect old or deleted pages to relevant new pages.
- **Monitor for Errors:** Use tools like Google Search Console to monitor 404 errors.

By implementing best practices for handling 404 errors and redirects, businesses can maintain their online authority, attract more traffic, and drive growth.

12. International SEO

If your website targets multiple countries or languages, implementing international SEO helps ensure the right content reaches the right audience.

- **Hreflang Tags:** Use hreflang tags to indicate language and regional targeting.

- **Localized Content:** Create content tailored to each target market.
  - Perhaps create a Spanish language page if you find one of your customer personas is Spanish-speaking.
- **Country-Specific Domains:** Consider using country-specific domains (e.g., .ca for Canada, .de for Germany).
  - <link rel="alternate" href="https://www.example.com" hreflang="en-us" />
  - <link rel="alternate" href="https://www.example.de" hreflang="de-de" />
  - <link rel="alternate" href="https://www.example.ca" hreflang="en-ca" />

More To Technical SEO Than You Realized

Technical SEO is not just about improving your website's search engine rankings—it's a strategic investment in your business's future growth. By optimizing every technical aspect of your site, you enhance user experience, which leads to higher engagement, longer visits, and more conversions. A well-optimized site builds trust and credibility, making users more likely to choose your services or products over competitors.

Investing in technical SEO is investing in the long-term success and profitability of your business. Providing a solid foundation that supports all your other marketing, leading to sustained growth and a stronger bottom line.

Chapter 8

ON-PAGE SEO BEST PRACTICES

Imagine stepping into a store where everything is meticulously organized, the aisles are clearly labeled, and every product is easy to find. This is the essence of effective on-page SEO. It's about creating a website that is not only user-friendly but also strategically optimized to attract and convert visitors into customers, ultimately driving higher ROI.

Success in growing your business online hinges on more than just having a visually appealing website or compelling content. It's about optimizing every aspect of your web pages with a clear, strategic approach.

On-page SEO encompasses all the elements you can control on your website to improve its search engine performance, from the content you produce to the HTML elements behind the scenes. There are overlaps with technical SEO, but the focus remains on optimizing individual pages.

Effective on-page SEO ensures your website ranks higher on search engine results pages (SERPs), making it easier for potential customers to find you. This increased visibility translates directly to more organic traffic and can complement your paid advertising efforts, thereby lowering the cost per lead and resulting in higher ROI.

The Importance of On-Page SEO

On-page SEO involves optimizing individual web pages to rank higher and earn more relevant traffic from search engines. It encompasses both the content and the HTML source code of a page.

Remember that the user always comes first. By creating a seamless, valuable experience for your users, you can naturally align with SEO best practices.

Title Tags

Title tags are HTML elements that specify the title of a web page. They are displayed on search engine results pages (SERPs) as clickable headlines for given results and are important for usability, SEO, and social sharing.

- **Include Primary Keywords:** Make sure your primary keyword appears in the title tag, preferably towards the beginning.
- **Keep It Concise:** Title tags should be between 50-60 characters to ensure they are fully displayed in SERPs.
- **Make It Compelling:** Write titles that encourage clicks by being clear and enticing.

User-First Approach: Think about what will catch the user's eye and make them want to click. What are they looking for? What problems are they trying to solve? Your title tag should speak directly to them.

Example

Original Title: "Water Bottles"

Optimized Title: "Buy Eco-Friendly Water Bottles - Sustainable & Reusable"

Meta Descriptions

Meta descriptions provide concise summaries of web pages. They appear under the clickable links in search engine results and can influence click-through rates.

- **Include Primary Keywords:** Keywords should be naturally included in the description.
- **Keep It Within 150-160 Characters:** Ensure the meta description is fully displayed in SERPs.
- **Be Persuasive:** Write descriptions that provide a compelling reason to click through to your site.

User-First Approach: Your meta description should address the user's needs and motivations. Why should they click on your link? What value will they get from visiting your page?

Example

Original Meta Description: "We sell water bottles."

Optimized Meta Description: "Discover our range of eco-friendly water bottles. Perfect for everyday use and environmentally friendly. Shop now and reduce plastic waste."

Header Tags (H1, H2, H3, etc.)

Header tags are HTML elements (H1, H2, H3, etc.) used to define headings and subheadings on a webpage. They help structure your content and make it easier for both users and search engines to understand.

- **H1 Tag:** This is the most important header tag and should be used to indicate the main topic of your page. Each page should have one H1 tag that includes the primary keyword and clearly communicates the page's primary focus. The H1 tag helps search engines understand the overall theme of your page.

- **H2 Tags:** These are used for main sections within your content. H2 tags help break your content into digestible parts, making it easier for readers to follow and understand. They also signal to search engines the major subtopics of your content.
- **H3 Tags:** These tags are used for subsections within H2 sections. They provide additional structure and detail, helping to further organize your content. H3 tags help both users and search engines understand the finer details within your main sections.

User-First Approach: Structure your content logically to enhance readability and user experience. Clear headings guide users through your content and help them find the information they need quickly.

Header Tag Example

An example of a well-structured page for an attorney might have an H1 tag like "Experienced Family Law Attorneys," an H2 tag such as "Why Choose Our Family Law Firm?" and an H3 tag like "Comprehensive Legal Services for Family Matters."

Content Quality and Relevance

High-quality content is informative, engaging, and relevant to your audience. It should address the needs and questions of your target audience, providing value and encouraging interaction. We will go more in-depth when we discuss content marketing.

- **Focus on User Intent:** Create content that satisfies the search intent of your target keywords.
- **Be Comprehensive:** Cover topics in-depth to provide valuable information.

- **Use Keywords Naturally:** Integrate keywords seamlessly into your content.

User-First Approach: Your content should be designed to answer the questions and solve the problems of your ideal customer. Provide valuable information that meets their needs and keeps them coming back for more.

Quality & Relevance Example

Instead of generic content like "Our attorneys are the best," optimized content would be "Looking for expert family law attorneys? Our team provides comprehensive legal services, including divorce, child custody, and spousal support. Learn how we can help protect your rights and interests."

Internal Linking

Internal linking involves linking to other pages within your website. It helps search engines understand the structure of your site and establishes a hierarchy of importance for your content.

- **Use Descriptive Anchor Text:** Use text that clearly describes the linked page.
- **Link to Relevant Content:** Ensure the linked content is relevant and provides additional value to the reader.
- **Distribute Link Equity:** Link to both high-priority (service pages) and supporting pages (blog posts) to distribute authority throughout your site.

User-First Approach: Internal links should enhance the user's journey by providing them with additional valuable content and improving site navigation.

Internal Linking Examples

Original Internal Link: "Click here"

Optimized Internal Link: "Benefits of our eco-friendly water bottles." Yes, the link is long but it will hold more relevance and link strength behind it.

- **Internal Linking:** Create a strong internal linking structure to help search engines discover your content. Internal links are hyperlinks that point from one page on your website to another page on the same site. They help search engines understand the relationship between different pages and prioritize which pages to crawl.
    - **Example 1:** Suppose you run an IT-managed services company. When you publish a new article on "The Benefits of Managed IT Services," you should include internal links to your previous articles such as "What is Managed IT Services?", "How to Choose the Right IT Service Provider," and "Top Cybersecurity Practices for Businesses." This not only helps readers find related content but also enables search engines to crawl and index these pages efficiently.
    - **Example 2:** On your local service pages, if you offer IT services in different cities, include links to specific service pages for each location. For example, if you have a page for "IT Services in San Fransisco," you should link to related local pages such as "IT Services in Orlando," "IT Services in Sacramento," and "IT Services in Stockton." This helps search engines understand the relationship between your service locations and improves local SEO
        - **Here's a tip:** you can link to other city service pages by just adding a section at the bottom of the page that lists other cities you service and linking the cities on that list to their respective pages.

The Reader Comes First

Your SEO strategy and implementation should always prioritize the user. Understanding your ideal customer is important to creating content that resonates with them and drives engagement. SEO is not just about pleasing Google; it's about providing real value to your users. Don't add links where it doesn't make sense just for the sake of adding them.

Image Optimization

Image optimization involves reducing the file size of images without compromising quality to ensure faster loading times. It also includes using descriptive file names and alt text to help search engines understand the content of the images.

Use Descriptive File Names:

File names are the first point of interaction between your images and search engines. Using descriptive, keyword-rich titles for your image files can significantly improve your SEO.

- **Avoid Generic Names:** Replace generic names like IMG001.jpg with descriptive names like red-ceramic-mug.jpg.
- **Incorporate Keywords:** If you're targeting specific keywords, include them in the file name. For instance, dallas-landscaping-service.jpg instead of landscaping.jpg.
- **Be Specific:** Describe the image clearly and specifically, e.g., blue-sports-car-2024-model.jpg rather than car.jpg.

Add Alt Text:

Alt text, or alternative text, is a vital part of image optimization. It serves as a textual description of an image, which is essential for both accessibility and SEO.

- **Descriptive and Relevant:** Make sure your alt text accurately describes the content and function of the image. For example, alt="Golden retriever puppy playing with a ball in the park" is better than alt="dog".
- **Incorporate Keywords Naturally:** Use relevant keywords in the alt text, but avoid keyword stuffing. A natural incorporation of keywords is beneficial for SEO. For example, alt="Professional landscaping service in Dallas, Texas" if it fits the image content.
- **Functional Descriptions:** If the image is functional (like a button or link), describe its function. For example, alt="Submit Form Button".

Example of Optimized Image Implementation

Consider a landscaping company website with the following image optimization:

- **File Name:** modern-backyard-garden-design.jpg
- **Alt Text:** alt="Modern backyard garden design with stone path and flower beds"
- **Compressed Image:** Compress the file size of the image to reduce load time.

By using these best practices, the image is more likely to appear in image search results, the page might load faster for users, and provide a better experience for those using screen readers.

Chapter 9

GOOGLE BUSINESS PROFILE

Imagine unlocking a secret weapon for your business that not only boosts your local SEO but also skyrockets your online visibility and drives a flood of traffic to your website—all for free. Picture this tool as a magnet, drawing local customers to your doorstep, and making your business the go-to option in your area. Sounds too good to be true?

Welcome to the game-changing world of Google Business Profile (GBP). This powerful tool has the potential to revolutionize your local marketing efforts, making your business more accessible and attractive to local customers.

Google Business Profile is more than just a simple business listing on Google. It's an interactive platform that allows you to engage with customers, showcase your products and services, and provide essential information that can lead to more conversions. Imagine being able to post updates about your latest promotions, respond to customer reviews in real-time, and gain valuable insights into how people are finding and interacting with your business. All of this is possible with GBP.

In this chapter, we'll embark on an in-depth exploration of Google Business Profile, uncovering its features, benefits, and the strategies you need to leverage it effectively. We'll delve into how GBP can transform your local SEO efforts, increase your online visibility, and ultimately drive more traffic and sales to your business.

Ready to unlock the full potential of your business? Let's dive into the world of Google Business Profile.

The Importance of a Google Business Profile

Google Business Profile is more than just a listing; it's a dynamic platform that connects your business with local customers.

- **Improved Local SEO:** GBP listings are a key factor in local search rankings. Optimizing your profile can help you appear in the coveted Google Local Pack, Maps, and local search results.
- **Increased Visibility:** A well-optimized profile can make your business stand out in search results, increasing your chances of attracting potential customers.
- **Enhanced Engagement:** GBP allows you to interact with customers through reviews, Q&A, and posts, fostering a stronger relationship with your audience.
- **Valuable Insights:** GBP provides analytics on how customers find and interact with your business, helping you refine your marketing strategies.

Setting Up and Verifying Your Google Business Profile

If you haven't already, the first step is to create and claim your Google Business Profile.

Google is consistently changing the way businesses are verified on Google Business Profile. Always check the latest guidelines on the Google Business Profile website for the most current verification process.

Steps to Verify the Google Business Profile

- **Go to Google Business Profile:** Visit the Google Business Profile website and sign in with your Google account.
- **Enter Your Business Information:** Provide your business name, address, and contact information. Ensure your details are accurate and consistent with other online listings.
- **Verify Your Business:** Google now uses video verification through the Google Business Profile app on your phone. Follow these steps to complete the video verification:

- o **Request Verification:** After entering your business details, select the option to verify your business and choose the video verification method.
- o **Prepare for the Video Call:** Ensure you have the Google Business Profile app installed on your smartphone.
- o **Record the Verification Video:** The video verification process should take about 2 minutes. You can take longer but the longer your video is the larger the file size and there is a possibility that it could be rejected when you upload it. Here's what to include:
  - **Street and Address:** Start by recording the street and the address of your business. If you work from home and the street name is too far away, show the house number from the street or driveway.
  - **Access to Building:** Record yourself accessing the building or home using your key or code to enter. This demonstrates that you have genuine access to the location.
  - **Business Interior:** Quickly go to your desk or business area while still recording.
  - **Business License:** Show your business license clearly, ensuring the state seal, business name, registered owner name, address, UIB number, and DBA name (if applicable) are visible.
  - **Google Business Profile Access:** Show on your computer that you have access to your Google Business Profile, demonstrating you can manage and edit the listing.
  - **Additional Proof (Optional):** While recording, you can also show your HR system or payroll dashboard, such as a Stripe account dashboard, to prove your business is active.

Make sure you are already logged in to avoid showing any passwords.

- Some companies suggest showing business merchandise or cards, but these don't prove ownership. It's better to show a business license or access to internal systems

- **Complete Verification:** Once you submit the video, a Google representative will review it and confirm your business profile. This can take anywhere from 3-30 days.

Why Google Uses Video Verification

Google has transitioned to video verification for several reasons:

- **Enhanced Security:** Video verification reduces the risk of fraudulent business listings. By visually confirming the business location and the owner's access, Google ensures that the profile represents a legitimate business.
- **Real-time Confirmation:** The video call allows Google to verify multiple aspects of the business in real-time, ensuring that the business is operational and matches the provided information.

Optimizing Your Google Business Profile

Add as much information as possible to your business profile to get the most out of it.

- **Business Name:** Use your official business name.
- **Address:** Ensure your address is correct and consistent across all platforms.
- **Phone Number:** Provide a phone number that connects directly to your business.
- **Website:** Include your website URL.
- **Business Hours:** List your operating hours accurately.

- **Category:** Choose the most relevant category for your business. This helps Google understand what your business offers and match it with relevant searches.
- **Attributes:** Highlight specific features such as "Free Wi-Fi" or "Wheelchair Accessible" if it applies to your business.

Add High-Quality Photos and Videos

Visual content can play a big role in engaging customers and showcasing your business.

- **Business Logo:** Upload your business logo.
- **Cover Photo:** Choose an appealing cover photo that represents your brand.
- **Interior and Exterior Photos:** Add photos of your business to give customers a sense of what to expect.
- **Product and Service Photos:** Showcase your products or services with high-quality images.
- **Videos:** Include short videos that highlight your business, services, or customer testimonials.

Utilize Google Posts

Google Posts allow you to share updates, offers, events, and more directly on your profile.

Action Steps:

- **Create Regular Posts:** Post updates about new products, services, promotions, or events.
- **Use High-Quality Images:** Ensure your posts are visually appealing.

- **Include CTAs:** Add clear calls-to-action such as "Learn More," "Call Now," or "Book Appointment."

Example:

- **Promotion Post:** "Get 20% off on all teeth whitening services this month! Book your appointment today."
- **Event Post:** "Join us for our free dental health workshop on April 15th. Reserve your spot now!"

Encourage and Manage Reviews

Reviews are critical for building trust and credibility. Encourage satisfied customers to leave positive reviews and respond to all reviews promptly.

- **Ask for Reviews:** After a successful interaction, ask customers to leave a review.
- **Respond to Reviews:** Thank customers for positive reviews and address any concerns in negative reviews professionally.

Responding to Negative Reviews

- **Stay Calm and Professional:** Responding in a heated manner can make things worse. Take time to cool off before responding.
- **Acknowledge the Issue:** Show empathy and understanding of the customer's concern.
- **Offer to Resolve Offline:** Suggest taking the conversation offline to resolve the issue. Provide contact details or invite the customer to call or email you directly.
- **Keep it Brief:** Avoid long explanations or justifications. Acknowledge, apologize, and offer a solution.

Example Response:

- **Positive Review:** "Thank you for your kind words! We're thrilled you had a great experience and look forward to seeing you again."
- **Negative Review:** "We're sorry to hear about your experience. Please contact us directly at [phone number/email] so we can resolve the issue and improve our services."

Use the Q&A Feature

The Q&A feature allows potential customers to ask questions about your business. Answer promptly and accurately to build trust and provide valuable information.

- **Monitor Questions:** Regularly check your profile for new questions.
- **Provide Detailed Answers:** Give clear and helpful responses to each question.
- **Add Common Questions:** Preemptively add and answer frequently asked questions to provide quick information.

Example Questions and Answers:

- **Q:** "Do you offer emergency dental services?"
- **A:** "Yes, we provide emergency dental services. Please call us at [phone number] for immediate assistance."

What Does All This Mean For Your Business?

First, having an optimized GBP boosts your local SEO, which is the key for being found by local customers. When your profile is well-managed, your business is more likely to appear in Google Maps and the Local Pack. These are the top spots in local search results that most users see and trust, making it easier for potential customers to discover your business.

Encouraging customers to leave reviews and responding to them—whether positive or negative—builds your reputation. Positive reviews attract new customers, while a well-handled negative review can show your commitment to customer satisfaction.

The insights provided by GBP are invaluable. You can see how customers find and interact with your profile, tracking metrics like the number of views and actions taken (such as calls or direction requests). This data helps you understand your customers better and refine your marketing strategies to meet their needs or solve their pain points more effectively.

An optimized GBP also gives you a competitive edge. If your profile is more detailed and engaging than your competitors, customers are more likely to choose you. It's a cost-effective way to stand out, as GBP is a free tool that can deliver substantial benefits with just an investment of time and effort.

Chapter 10

OFF-PAGE SEO BEYOND THE PAGE

There is an elusive force, a hidden power that could elevate your website's ranking and authority without altering a single element on your site. This mysterious and potent force is known as Off-Page SEO. It operates in the shadows, beyond the borders of your website, yet its impact is unmistakable and profound. Imagine your website as a ship sailing through the vast ocean of the internet.

On-page SEO ensures your ship is well-built and optimized for the journey. However, off-page SEO acts as the favorable winds and currents, propelling your ship toward its destination faster and more efficiently.

We'll uncover the secrets of Off-Page SEO, exploring its importance, strategies, and how you can leverage it to grow your business.

The Importance of Off-Page SEO

Off-Page SEO involves building your site's authority, credibility, and relevance through external means. Here's why you need it and why many marketing companies don't utilize it. (spoiler: it's difficult to do correctly)

- **Enhanced Authority:** High-quality backlinks from reputable sites signal to search engines that your site is trustworthy. Building such a profile requires a strategic approach and creating valuable content that others want to link to.
- **Improved Rankings:** Search engines consider off-page factors, such as backlinks, social signals, and brand mentions, when

determining your site's ranking. Unlike on-page SEO, where you have direct control over your site's elements, off-page SEO involves influencing external factors, making it a more complex and challenging endeavor.

- **Increased Traffic:** Effective off-page strategies can drive more referral traffic from external sites and social media platforms.
- **Greater Brand Visibility:** Off-page activities help increase your brand's exposure and recognition across the web.

Understanding High-Authority Backlinks

High-authority backlinks are links from websites that are considered reputable, trustworthy, and authoritative by search engines. These backlinks carry more weight and can significantly boost your site's search engine rankings and credibility.

Characteristics of High-Authority Backlinks

- **Domain Authority (DA):** Websites with a high domain authority (DA), as measured by tools like Moz, are seen as more trustworthy and influential.
- **Relevance:** Links from websites that are relevant to your industry or niche are more valuable. They indicate to search engines that your content is relevant to the topic.
- **Traffic:** Sites with high traffic indicate that they are popular and trustworthy, making backlinks from these sites more beneficial.
- **Editorial Links:** These are links given naturally by other websites, not through paid or reciprocal arrangements. Editorially placed links are a strong endorsement of your content.
- **Content Quality:** Backlinks from sites known for high-quality content are more valuable. These sites typically have stringent editorial standards and produce valuable, in-depth content.

- **Trustworthiness:** Links from government (.gov), educational (.edu), and major news websites are considered highly authoritative due to their trustworthiness.

Sources of High-Authority Backlinks

- **Guest Blogging:** Contributing high-quality guest posts to authoritative blogs in your industry can earn you valuable backlinks.
- **Industry Publications:** Getting featured in respected industry magazines, journals, or blogs.
- **Media Coverage:** Press mentions and news articles about your business can provide powerful backlinks.
- **Influencer Endorsements:** When well-known industry influencers link to your content, it enhances your site's credibility.
- **Educational Institutions:** Collaborations, studies, or projects with universities and colleges can result in authoritative backlinks.
- **Government Websites:** Participating in government programs or being listed as a resource on government websites can yield highly authoritative links.
- **Resource Pages:** Getting listed on resource pages or directories relevant to your field.
- **Partnerships and Sponsorships:** Collaborating with other reputable businesses or sponsoring events can lead to high-quality backlinks.

Why High-Authority Backlinks Matter

- **Improved Rankings:** Search engines use backlinks as a signal of the quality and relevance of a site. High-authority backlinks are more likely to improve your search engine rankings.
- **Increased Traffic:** These links can drive referral traffic from authoritative sites to your website.

- **Enhanced Credibility:** Being linked to by reputable sites enhances your site's credibility and trustworthiness in the eyes of both search engines and users.
- **Networking Opportunities:** Building relationships with high-authority sites can lead to more collaboration opportunities and long-term benefits.

Beware of Buying Links

In the pursuit of high-quality backlinks, it can be tempting to take shortcuts, such as buying links. This might seem like an easy way to quickly build a robust backlink profile, but it's fraught with risks and pitfalls.

Why Buying Links is Risky

- **Quality vs. Quantity:** Purchased links often come from low-quality, irrelevant sites. Even if you get thousands of links, they might not improve your rankings and could even harm your site's reputation.
- **Search Engine Penalties:** Search engines like Google have strict policies against buying links. If they detect that you've bought links, your site could be penalized, leading to a significant drop in rankings or even removal from search results.
- **Scams and False Promises:** Many services that sell links make grand promises but fail to deliver. They might provide links that disappear after a short period or come from spammy websites.
- **Lack of Control:** When you buy links, you have no control over the quality and relevance of the sites linking to you. This lack of control can lead to poor-quality links that offer no real value. The sites might say they provide high profile links but to be frank. They are more often than not, lying.

The Temptation and the Reality

Buying links can be very tempting because it appears to be a quick fix. The idea of instantly gaining thousands of backlinks is alluring, especially when you're trying to compete in a crowded market.

However, the easy way is rarely the most profitable way in the long run. Genuine, high-quality backlinks require effort, time, and strategy, but they yield sustainable results that far outweigh the temporary gains from purchased links.

Expert Advice

Instead of buying links, invest in creating high-quality content that naturally attracts backlinks. Build relationships with industry influencers and focus on earning editorial links from reputable sites. This approach, though more challenging, is far more rewarding and aligns with search engine guidelines.

Key Off-Page SEO Strategies

Building High-Quality Backlinks

Backlinks are one of the most critical factors for off-page SEO. They act as votes of confidence from other websites, indicating that your content is valuable and worth linking to.

- **Guest Blogging:** Write high-quality guest posts for reputable websites in your industry. Include a link back to your site in the author bio or content where appropriate.
  - **Example:** A moving company can write a guest post on a popular real estate blog about tips for a smooth relocation, including a link back to their site.
- **Content Promotion:** Share your content with influencers and industry leaders who might find it valuable enough to link to.

- **Example:** A moving company can create a detailed guide on the best practices for moving during the winter and share it with home improvement influencers.
- **Broken Link Building:** Find broken links on other websites and suggest your content as a replacement.
  - **Example:** A moving company can identify broken links in home organization resource lists and offer their own moving checklists as replacements.
- **Resource Pages:** Get your site listed on resource pages or directories relevant to your industry.
  - **Example:** A moving company can get listed on directories that compile moving resources and services for new homeowners.

Social Media Engagement

Social media platforms are powerful tools for promoting your content and engaging with your audience.

- **Share Content Regularly:** Post your blog articles, promotions, and updates on your social media channels.
  - **Example:** A moving company can regularly post packing tips, customer testimonials, and promotional offers on their social media accounts to keep their audience engaged.
- **Engage with Followers:** Respond to comments, answer questions, and participate in discussions to build a community around your brand.
  - **Example:** A moving company can respond to customer inquiries and comments, providing personalized moving advice and engaging in related discussions.
- **Collaborate with Influencers:** Partner with influencers in your industry to expand your reach and credibility.
  - **Example:** A moving company can collaborate with popular home organization bloggers or real estate influencers to share moving tips and recommendations.

- **Run Social Media Campaigns:** Create campaigns to promote special offers, events, or new content.
  - **Example:** A moving company can run a social media campaign offering discounts on moves booked during off-peak seasons, encouraging users to share their moving stories for a chance to win a prize.

Online Reviews and Reputation Management

Online reviews are very important for building trust and credibility with potential customers.

- **Encourage Reviews:** Ask satisfied customers to leave reviews on Google, Yelp, and other review platforms.
  - **Example:** A moving company can encourage clients to leave reviews on Google and Yelp by offering a discount on their next service for every review left.
- **Monitor Reviews:** Regularly check review sites for new reviews and respond promptly.
  - **Example:** A moving company can assign a team member to monitor reviews on Yelp and respond to both positive and negative feedback.
- **Address Negative Reviews:** Respond professionally to negative reviews, addressing the issue and offering a solution.
  - **Example:** A moving company can respond to a negative review about a delayed move by apologizing, explaining the reason for the delay, and offering a discount on the next service.
- **Showcase Positive Reviews:** Highlight positive reviews on your website and social media.
  - **Example:** A moving company can feature glowing customer testimonials on their homepage and share these reviews on their social media channels.

While we highlight the role of online reviews in enhancing user experience and trust, a dedicated chapter will explore review management in depth. This will cover ethical solicitation, feedback response, and leveraging reviews to boost SEO and customer trust.

Local SEO and Citations

Local SEO is crucial for businesses that operate in specific geographic areas. Citations, or mentions of your business's name, address, and phone number (NAP) on other websites, are key for local SEO.

- **Claim and Optimize Listings:** Ensure your business is listed on major local directories like Google My Business, Yelp, and Bing Places.
  - **Example:** A local moving company can claim and optimize their Google My Business profile, including high-quality photos, business hours, and contact information.
- **Consistent NAP:** Ensure your business name, address, and phone number are consistent across all listings.
  - **Example:** A moving company can verify that their NAP information is consistent across their website, social media profiles, and local directory listings.
- **Get Listed in Local Directories:** Submit your business to local directories and industry-specific sites.
  - **Example:** A moving company can get listed in local business directories and home services-related sites.
- **Participate in Local Events:** Sponsor local events, join local business associations, and get featured in local news.
  - **Example:** A moving company can sponsor a local community event, gaining exposure in local news and community websites.

Content Marketing and Syndication

Content marketing involves creating valuable content and distributing it across various platforms to attract and engage your target audience.

- **Create High-Quality Content:** Develop informative and engaging content such as blog posts, infographics, and videos.
  - **Example:** A moving company can create detailed guides on packing, moving tips, and settling into a new home, complete with infographics and videos.
- **Syndicate Content:** Share your content on platforms like Medium, LinkedIn, and industry-specific forums.
  - **Example:** A moving company can publish articles on Medium and LinkedIn about the best practices for a stress-free move, reaching a broader audience.
- **Repurpose Content:** Convert blog posts into different formats, such as podcasts, videos, or infographics, and share them on relevant platforms.
  - **Example:** A moving company can turn a written moving checklist into a video series and share it on YouTube and social media.

Influencer and Community Engagement

Engaging with influencers and participating in community activities can boost your brand's visibility and credibility.

- **Identify Industry Influencers:** Find influencers in your industry who have a large and engaged following.
  - **Example:** A moving company can identify popular home organization and real estate influencers who align with their brand values and audience.
- **Collaborate with Influencers:** Partner with influencers for guest posts, social media takeovers, or joint events.
  - **Example:** A moving company can collaborate with a home organization influencer for a social media takeover, showcasing packing and moving tips.
- **Join Industry Communities:** Participate in industry-specific forums, groups, and online communities.
  - **Example:** A moving company can join forums and LinkedIn groups related to home services and relocation to share insights and network with other professionals.

- **Host Webinars and Workshops:** Organize online or offline events to share your expertise and engage with your audience.
  - o **Example:** A moving company can host webinars on moving tips and strategies, inviting industry experts to speak and engage with the audience.

Measuring Off-Page SEO Success

Tracking the performance of your off-page SEO efforts is crucial for understanding their impact and making necessary adjustments.

- **Backlink Profile:** Use tools like Ahrefs or SEMrush to monitor the number and quality of backlinks to your site.
  - o **Example:** Monitor the growth of backlinks to your site over time and analyze the domain authority of the linking sites.
- **Referral Traffic:** Track the amount of traffic coming from external sites using Google Analytics.
  - o **Example:** Identify which external sites are driving the most traffic to your site and optimize your strategies accordingly.
- **Social Engagement:** Measure likes, shares, comments, and overall engagement on your social media posts.
  - o **Example:** Track the performance of different types of social media content to determine what resonates most with your audience.
- **Review Ratings:** Monitor the number and quality of reviews on platforms like Google and Yelp.
  - o **Example:** Regularly review your ratings and feedback to identify areas for improvement and showcase positive reviews.
- **Local Search Rankings:** Check your rankings for local search terms in tools like Google Search Console and BrightLocal.
  - o **Example:** Track your rankings for key local search terms and optimize your local SEO strategies based on performance data.

Real-World Example: Move With Ease

- **Building High-Quality Backlinks:** Move With Ease writes a guest post on a popular real estate blog about tips for a smooth relocation, including a link back to their site.
- **Social Media Engagement:** They share packing tips, customer testimonials, and special promotions on their Facebook and Instagram pages, engaging with their audience through comments and messages.
- **Online Reviews and Reputation Management:** After each move, the company asks satisfied clients to leave reviews on Google and Yelp, responding to each review to show appreciation and address any concerns.
- **Local SEO and Citations:** The company ensures their Google My Business profile is fully optimized and gets listed in local directories like HomeAdvisor and Angie's List.
- **Content Marketing and Syndication:** They create a series of blog posts on moving tips, repurpose the content into infographics, and share them on social media and home improvement forums.
- **Influencer and Community Engagement:** The company collaborates with a local home organization influencer to promote a stress-free moving campaign and hosts a free community workshop on moving tips.

The Power of Quality Over Quantity

In the realm of off-page SEO, the mantra "quality over quantity" holds significant power. Not all backlinks are created equal, and the value of a few high-quality links from reputable, authoritative sites far surpasses that of numerous low-quality links.

High-authority backlinks act as strong endorsements from trusted sources, signaling to search engines that your content is credible and valuable. This not only enhances your site's authority but also boosts your search engine rankings more effectively.

Investing time and effort into securing high-quality backlinks from relevant industry sites can yield substantial long-term benefits. It's about building a network of trust and authority, rather than merely amassing links.

Quality links drive more meaningful referral traffic, enhance your brand's reputation, and ultimately lead to sustainable growth in your online visibility. Focus on creating valuable content that attracts these high-quality links naturally, and prioritize genuine relationships within your industry to maximize your off-page SEO efforts.

Implement these off-page SEO strategies to elevate your business and watch your online visibility soar. Ready to take your SEO efforts to the next level? Let's dive into the detailed steps and tactics that will help you master Off-Page SEO.

Chapter 11

GROWING YOUR ONLINE REVIEWS

Imagine walking into a bustling marketplace where word of mouth spreads faster than any advertisement. Online reviews are the modern-day word of mouth, wielding immense power in shaping your business's reputation and influencing potential customers.

Contrary to the belief that online reviews are fake, genuine feedback from real customers can significantly impact your business. Positive reviews can enhance your credibility, attract more customers, and even improve your search engine rankings.

This chapter will guide you through the nuances of managing online reviews effectively, from soliciting authentic feedback to responding appropriately and leveraging reviews to boost your SEO, ultimately driving your business growth.

Do Online Reviews Matter?

Online reviews are not just a trend; they are a cornerstone of digital credibility. Here's why they matter:

- **Building Trust:** Just like personal recommendations, online reviews help build trust with potential customers. Positive reviews signal that your business is reliable and delivers quality service.
- **Boosting SEO:** Google values the quantity and quality of reviews when ranking websites. A robust review profile can significantly improve your search engine visibility.
- **Increasing Conversions:** Positive reviews can tip the scales for undecided customers, making them more likely to choose your business over competitors.

- **Gathering Insights:** Reviews offer direct feedback from your customers, providing valuable insights into what you're doing right and areas that need improvement.

Navigating Google's Guidelines for Reviews

To maintain the integrity of online reviews, Google has set clear guidelines. Not following these rules could put your Google Business Profile in jeopardy.

- **Offer Incentives:** Offering rewards like discounts or freebies in exchange for reviews is prohibited. Such practices can lead to biased and unreliable feedback.
- **Ask for Positive Reviews Only:** Soliciting only positive feedback skews the authenticity of reviews. Ask all customers for honest reviews.
- **Review Gating:** Filtering customers and directing only satisfied ones to leave reviews is against Google's policies. Every customer should have an equal chance to leave feedback.
- **Bulk Requests:** Avoid sending mass emails to solicit reviews. Google prefers a steady and natural flow of reviews over time.
- **Review Swapping:** Exchanging reviews with other businesses can lead to biased feedback and is against the guidelines.
- **Fake Reviews:** Creating fake reviews or having employees post reviews is a clear violation.
- **Third-Party Services:** Using services that generate positive reviews for you can result in severe penalties such as removal of those reviews and even possibly your site being removed from Google altogether.

Building Genuine Reviews

Rather than resorting to unethical practices, focus on building a robust strategy to garner genuine reviews.

- **Provide Excellent Service:** The foundation of positive reviews is excellent service. Strive to exceed customer expectations in every interaction.
- **Ask at the Right Time:** Request reviews when customers have had a positive interaction or completed a successful service. For example, ask for a review immediately after a successful move if you own a moving company.
- **Make It Easy:** Provide clear instructions and direct links to your review platforms. Simplify the process as much as possible for your customers.

Effective Methods for Soliciting Reviews

- **Email Campaigns:** Follow up with customers through email, thanking them for their business and politely asking for their feedback. Include a direct link to your Google review page.

Example:

Hi [Customer Name],

Thank you for choosing [Your Company] for your recent move. We hope you had a great experience. We would appreciate it if you could take a moment to leave us a review on Google. Your feedback helps us improve and helps others make informed decisions.

[Insert Link to Google Review Page]

- **Follow-Up Calls:** After a service, follow up with a phone call to ensure the customer is satisfied and kindly ask if they would leave a review.

- **In-Person Requests:** For businesses with face-to-face interactions, ask for reviews in person at the end of a service. Provide a card with instructions on how to leave a review.

How to Respond to Reviews

- **Positive Reviews:** Thank the customer and acknowledge their feedback. Personalize your response to show appreciation.

Example:

Hi [Customer Name],

Thank you for your kind words! We're thrilled to hear you had a great experience with us. We look forward to serving you again.

- **Negative Reviews:** Address the issue professionally and offer a solution. Show that you take their concerns seriously and are willing to make things right.

Example:

Hi [Customer Name],

We're sorry to hear about your experience. This is not the standard of service we aim to provide. Please contact us at [Your Contact Information] so we can address your concerns and make it right.

- **Turning Negative Reviews into Positive Opportunities**

Use negative reviews as opportunities to improve your services. Addressing concerns publicly shows potential customers that you are responsive and committed to quality.

Leveraging Reviews for Local SEO

Importance of NAP Consistency

Ensure your business's Name, Address, and Phone number (NAP) are consistent across all review platforms and local directories.

Using Reviews to Enhance Local Listings

Positive reviews can boost your local search rankings. Encourage satisfied customers to mention specific services and locations in their reviews.

Getting Featured in Local Directories

Submit your business to local directories and ensure your NAP information is accurate. Positive reviews in these directories can enhance your local SEO.

Online reviews are a huge step to build your site's authority, improving search engine rankings, and driving more traffic.

By following best practices for soliciting and managing reviews, you can significantly enhance your online presence and attract more customers. Remember, quality over quantity is key.

Focus on providing excellent service and encouraging genuine feedback to build a strong, trustworthy online reputation.

Chapter 12

STRATEGIC CONTENT MARKETING

Imagine if you could create content that not only attracts your ideal customers but also keeps them engaged and drives them to take action. What if your blog posts, videos, and infographics could position your business as an industry leader, build trust with your audience, and ultimately boost your bottom line? This is the power of content marketing.

Much like the previous chapters on understanding your ideal customer, analyzing competitors, and mastering SEO, content marketing is a crucial piece of the puzzle that can significantly enhance your digital marketing strategy.

Content marketing isn't just about creating random pieces of blogs or service pages; it's about developing a strategic approach that aligns with your business goals and resonates with your audience. This chapter will provide you with a deep, strategic understanding of content marketing, offering valuable insights and actionable tactics to help you create and leverage content that drives business growth.

We'll start by defining the key elements of a successful content marketing strategy, from understanding your audience to conducting keyword research and creating a content plan. Then, we'll explore how to develop high-quality content, promote it effectively, and measure its success. Along the way, we'll provide tangible examples and real-world case studies to illustrate how these strategies can be applied to drive business growth.

Whether you're a seasoned business owner or just starting out, this chapter will equip you with the knowledge and tools you need to harness the power of content marketing and elevate your business. Ready to dive in? Let's get started.

Bringing It All Together

In previous chapters, we've thoroughly explored how to understand your ideal customer, analyze your competitors, conduct keyword research, and master both on-page and technical SEO. Now, we will bring all these elements together to create a comprehensive content marketing strategy.

By leveraging your insights from each of these areas, you can develop content that not only attracts but also engages and converts your target audience.

Understanding Content Marketing

Content marketing is a strategic approach focused on creating and distributing valuable, relevant, and consistent content to attract and retain a clearly defined audience—and, ultimately, to drive profitable customer action.

Unlike traditional advertising, content marketing doesn't explicitly promote a brand but is intended to stimulate interest in its products or services through valuable information.

Why Content is King

Content is King. You've likely heard this phrase countless times, but what does it truly mean for your business? In essence, it signifies that high-quality, relevant content is the cornerstone of successful digital marketing. It's not just about creating content; it's about creating content that resonates with your audience, answers their questions, and solves their problems.

The phrase "Content is King" means that high-quality content can significantly influence your audience's engagement, trust, and purchasing decisions. Here's why:

- **Engagement:** Quality content engages your audience, keeping them on your site longer and increasing the likelihood of conversions. For example, an engaging blog post on "Top 10 Eco-Friendly Home Tips" can keep readers interested and encourage them to explore related products on your site.
- **Trust and Authority:** Providing valuable information builds trust and positions your brand as an authority in your industry. A comprehensive guide on "Understanding Solar Panel Installation" can position a solar energy company as an expert.
- **SEO Benefits:** Well-crafted content improves your search engine rankings by addressing user intent and incorporating relevant keywords naturally. For instance, a page optimized for "best organic skincare products" can attract targeted traffic from users searching for organic skincare solutions.
- **Customer Retention:** Consistent, valuable content helps retain customers by continuously meeting their needs and keeping your brand top-of-mind. Regular updates on industry trends and tips can keep an audience engaged and loyal.

Developing a Content Marketing Strategy

Creating a successful content marketing strategy involves several key steps:

- **Define Your Goals**
- **Understand Your Audience**
- **Conduct Keyword Research**
- **Create a Content Plan**
- **Develop High-Quality Content**
- **Promote Your Content**
- **Measure and Adjust**

1. Define Your Goals

Before you start creating content, it's crucial to define what you want to achieve. Your goals will guide your content strategy and help you measure success. Common content marketing goals include:

- **Increasing Brand Awareness:** Aim to reach new audiences and make your brand more recognizable.
- **Driving Website Traffic:** Create content that attracts visitors to your website.
- **Generating Leads:** Develop content that captures contact information from potential customers.
- **Nurturing Leads:** Use content to guide potential customers through the sales funnel.
- **Converting Leads into Customers:** Produce content that encourages final purchasing decisions.
- **Enhancing Customer Retention:** Keep existing customers engaged and loyal with valuable content.
- **Building Brand Authority:** Establish your brand as a leader in your industry through authoritative content.

2. Understand Your Audience

Understanding your audience is the foundation of effective content marketing. This involves creating detailed customer personas that represent your ideal customers, based on the insights you've gathered in previous chapters. Consider factors such as:

- **Demographics:** Age, gender, income, education, occupation.
- **Psychographics:** Interests, values, lifestyle, personality traits.
- **Goals:** What they aim to achieve with your product or service.
- **Pain Points:** Challenges or problems they face that your product or service can solve.

Example: Creating a Customer Persona

- **Persona:** Concerned Cathy

- o **Age:** 35
- o **Gender:** Female
- o **Location:** Urban area
- o **Occupation:** Healthcare professional
- o **Interests:** Pet care, health and wellness, organic products
- o **Goals:** Wants to find a reliable dog food that won't trigger her dog's allergies.
- o **Pain Points:** Struggles with finding affordable hypoallergenic dog food that her dog enjoys eating.

Conduct surveys, interviews, and social media analysis to gather insights about your audience. Use tools like Google Analytics to analyze the behavior and preferences of your website visitors.

3. Conduct Keyword Research

Keyword research is essential for understanding what your audience is searching for and how you can address their needs. Here's how to do it:

- **Brainstorm Seed Keywords:** Start with broad topics related to your business. For example, if you run a hypoallergenic dog food company, seed keywords might include "dog food allergies," "hypoallergenic dog food," or "dog food for sensitive stomachs."
- **Use Keyword Research Tools:** Tools like Google Keyword Planner, Ahrefs, and SEMrush can help you expand your list of potential keywords. For instance, using Google Keyword Planner, you might discover high-volume keywords like "best hypoallergenic dog food" or "natural dog food for allergies."
- **Analyze Keyword Metrics:** Look at search volume, keyword difficulty, and competition to prioritize keywords that are valuable and achievable. For example, "best hypoallergenic dog food" might have high search volume but medium difficulty, making it a good target.

- **Identify Long-Tail Keywords:** Focus on specific, intent-driven phrases that are more likely to convert, such as "best dog food for allergies" or "grain-free hypoallergenic dog food."

Example: Using Ahrefs for Keyword Research

- **Step 1:** Enter a seed keyword like "hypoallergenic dog food" into Ahrefs Keyword Explorer.
- **Step 2:** Review the keyword suggestions and filter by search volume and keyword difficulty.
- **Step 3:** Identify long-tail keywords such as "best hypoallergenic dog food for small breeds" or "natural hypoallergenic dog food recipes."

4. Create a Content Plan

A content plan outlines what content you will create, when you will publish it, and how it will support your marketing goals. Key elements of a content plan include:

- **Content Types:** Blog posts, videos, infographics, eBooks, case studies, webinars, podcasts.
- **Content Calendar:** Schedule for creating and publishing content to ensure consistency. For example, plan to publish a new blog post every Tuesday and a video tutorial every Thursday.
- **Topics:** List of topics that align with your audience's interests and search queries. For instance, a series of blog posts on "Managing Dog Allergies" or "Best Practices for Feeding Dogs with Food Sensitivities."
- **Distribution Channels:** Platforms where you will share your content, such as your website, social media, email newsletters, and guest blogs.

Example: Creating a Content Calendar

- **Month 1:**

- **Week 1:** Blog post on "Top 5 Signs Your Dog Has a Food Allergy"
- **Week 2:** Video tutorial on "How to Transition Your Dog to Hypoallergenic Food"
- **Week 3:** Infographic on "Common Dog Food Allergens to Avoid"
- **Week 4:** Guest blog post on a popular pet care website

Example: Creating a Blog-Only Content Calendar

Many companies don't have access to or the funds to create videos, infographics, or guest blog posts, so here is an example of effective blog content:

- **Month 1**
  - **Week 1:** Hypoallergenic Treats: Safe Snacking for Sensitive Dogs
    - **Content:** Recommend hypoallergenic treats for dogs with food sensitivities. Include product reviews and homemade treat recipes.
    - **Keywords:** hypoallergenic dog treats, dog treats for allergies, homemade hypoallergenic dog treats
  - **Week 2:** Understanding the Causes of Dog Food Allergies
    - **Content:** Provide an in-depth look at the causes of dog food allergies. Discuss genetic factors, environmental triggers, and common allergens.
    - **Keywords:** causes of dog food allergies, dog allergy causes, understanding dog food allergies
  - **Week 3:** How to Perform an Elimination Diet for Your Dog
    - **Content:** Guide readers through the process of conducting an elimination diet to identify food allergies in dogs. Include step-by-step instructions and tips for success.
    - **Keywords:** dog elimination diet, identifying dog food allergies, elimination diet for dogs
  - **Week 4:** The Impact of Nutrition on Canine Immune Health

- **Content:** Discuss how proper nutrition can boost a dog's immune system. Highlight the benefits of hypoallergenic diets for overall health and well-being.
- **Keywords:** canine immune health, nutrition and dog health, hypoallergenic diet benefits

This blog-only content calendar ensures a consistent flow of valuable information, addressing various aspects of hypoallergenic dog food and related topics. It keeps your audience engaged and informed while supporting your content marketing goals.

5. Develop High-Quality Content

Creating high-quality content involves several best practices:

- **Focus on User Intent:** Ensure your content addresses the search intent of your target keywords. For example, a blog post titled "How to Choose the Best Hypoallergenic Dog Food" should provide detailed, practical advice and product recommendations.
- **Be Comprehensive:** Cover topics in-depth to provide valuable information. An article on "The Complete Guide to Managing Dog Allergies" should explore various aspects like identifying symptoms, common allergens, and dietary solutions.
- **Use Keywords Naturally:** Integrate keywords seamlessly into your content without overstuffing. For instance, "hypoallergenic dog food" should appear naturally within the text, headers, and meta descriptions.
- **Engage Your Audience:** Use a mix of text, images, videos, and other media to keep readers engaged. A blog post could include step-by-step photos, an embedded video tutorial, and downloadable checklists.
- **Maintain a Consistent Tone:** Ensure your content reflects your brand's voice and tone. Whether it's professional, friendly, or authoritative, consistency is key.

Example: Creating an Engaging Blog Post

- **Title:** "10 Easy Ways to Manage Your Dog's Allergies"
- **Introduction:** Briefly explain why managing dog allergies is important.
- **Body:** Breakdown the 10 tips with detailed explanations, images, and examples.
- **Conclusion:** Summarize the key points and include a call-to-action to explore related hypoallergenic dog food products or subscription packages.

6. Promote Your Content

Creating great content is just the first step; you also need to promote it to reach a wider audience. Effective content promotion strategies include:

- **Social Media Marketing:** Share your content on social platforms where your audience is active. For example, promote your latest blog post on Facebook, Instagram, and LinkedIn with engaging visuals and captions.
- **Email Marketing:** Distribute content through newsletters and email campaigns to engage your subscribers. Include a mix of blog highlights, upcoming events, and exclusive offers.
- **Influencer Collaborations:** Partner with influencers to expand your reach and credibility. For instance, send samples of your hypoallergenic dog food to pet care influencers and ask them to share their reviews with their followers.
- **Guest Blogging:** Write guest posts for reputable sites in your industry to attract new audiences. Include links back to your own content and website.
- **Paid Promotion:** Use paid ads on social media and search engines to boost your content's visibility. Run targeted campaigns for high-performing blog posts or new product launches.

Example: Promoting a Blog Post

- **Step 1:** Share the blog post on your social media channels with a compelling image and caption.
- **Step 2:** Send an email newsletter to your subscribers highlighting the new blog post and why it's worth reading.
- **Step 3:** Reach out to influencers in your niche and ask them to share the post with their audience.
- **Step 4:** Use Facebook Ads to promote the blog post to a targeted audience interested in hypoallergenic dog food.

7. Measure and Adjust

Measuring the performance of your content is crucial for understanding what works and what doesn't. Use tools like Google Analytics, SEMrush, and social media insights to track key metrics such as:

- **Traffic:** Number of visitors to your content.
- **Engagement:** Time spent on page, bounce rate, social shares, comments.
- **Leads and Conversions:** Number of leads generated and conversion rates.
- **SEO Performance:** Keyword rankings and organic search traffic.

Example: Using Google Analytics

- **Step 1:** Set up goals in Google Analytics to track conversions.
- **Step 2:** Monitor the performance of individual blog posts and landing pages.
- **Step 3:** Analyze traffic sources to see where your visitors are coming from.
- **Step 4:** Adjust your content strategy based on performance data. For instance, if blog posts about "Managing Dog Allergies" are performing well, create more content on similar topics.

Selecting Valuable Content Topics

Choosing the right topics is critical for engaging your audience and meeting your marketing goals. Here are some strategies for selecting valuable content topics:

- **Address Customer Pain Points:** Identify common challenges your audience faces and create content that provides solutions. For example, "How to Identify and Treat Dog Food Allergies" addresses a common pain point for dog owners.
- **Leverage Keyword Research:** Use your keyword research to identify topics that align with your audience's search queries. If "best hypoallergenic dog food" is a popular search query, create a blog series on that topic.
- **Analyze Competitor Content:** Look at what your competitors are writing about and find opportunities to provide more value or a different perspective. This is a great way to find good content topics that could resonate with your target audience and that Google will like. Google favors relevant and updated content. Here's how to do it effectively:
  - **Evaluate Competitor Blogs and Articles:** Start by examining your competitors' blogs and articles. Identify the topics they cover, the depth of their content, and how frequently they publish. For instance, if your competitors have basic guides on dog allergies, assess how comprehensive these guides are. Look for gaps in information, outdated advice, or areas that could be expanded upon.
  - **Create More In-Depth Content:** If you find that your competitors only provide superficial information on a topic like dog allergies, you can create a more in-depth, step-by-step guide on managing these allergies. This might include detailed sections on identifying symptoms, common allergens, and specific dietary recommendations, complete with visuals and real-life examples.
  - **Provide a Different Perspective:** Offer a unique angle or perspective that your competitors haven't covered. For

example, if competitors focus on the general symptoms of dog allergies, you could create content that explores hypoallergenic diets for specific breeds or life stages, such as puppies versus senior dogs.

- o **Update and Improve Outdated Content:** Identify outdated content from competitors that still ranks well but lacks current information. Rewrite this content to make it more comprehensive, accurate, and up-to-date. For example, if you find a popular article on dog food allergies from a few years ago, update it with the latest research, new hypoallergenic products, and current best practices.
- o **Enhance Visual and Interactive Elements:** Improve upon competitors' content by adding high-quality images, infographics, videos, and interactive elements like quizzes or calculators. For instance, an interactive quiz to help dog owners determine if their pet might have a food allergy can be more engaging than a simple text article.
- o **Utilize User Feedback:** Incorporate feedback from your audience to refine and improve your content. Monitor comments on competitors' blogs and social media to understand what readers liked or found lacking. Use this information to enhance your content, ensuring it better meets the needs and interests of your audience.
- **Monitor Industry Trends:** Stay updated on trends and news in your industry to create timely and relevant content. For example, if there's a growing interest in organic pet food, publish articles and videos on the benefits of organic hypoallergenic dog food.
- **Engage with Your Audience:** Ask your audience directly through surveys, social media, and customer feedback to find out what topics they are interested in. Use tools like SurveyMonkey or simple Instagram polls to gather insights.

Example: Analyzing Competitor Content

- **Step 1:** Identify competitors in your niche and list their top-performing content pieces.

- **Step 2:** Evaluate the quality, depth, and relevance of their content. Look for gaps or opportunities for improvement.
- **Step 3:** Create a content plan to develop more comprehensive, engaging, and up-to-date articles. For example, if a competitor's guide on dog allergies is basic, plan to write an in-depth guide covering everything from symptoms and diagnosis to treatment options and preventive measures.
- **Step 4:** Add unique value by incorporating visuals, real-life examples, and interactive elements. Update your content regularly to keep it fresh and relevant.

Example Implementation:

- **Original Competitor Content:** "Top 5 Signs Your Dog Has a Food Allergy"
- **Your Enhanced Content:**
  - **Title:** "The Complete Guide to Identifying and Managing Dog Food Allergies"
  - **Introduction:** Briefly introduce the topic and explain the importance of managing dog food allergies.
  - **Sections:**
    - **Signs and Symptoms:** Detailed descriptions with images.
    - **Common Allergens:** List and describe common allergens found in dog food.
    - **Diagnosis:** Step-by-step guide on how to diagnose food allergies in dogs, including vet advice and home tests.
    - **Treatment Options:** Various treatment options, including hypoallergenic diets, medications, and home remedies.
    - **Preventive Measures:** Tips on how to prevent future allergic reactions.
    - **Interactive Quiz:** "Is Your Dog Suffering from a Food Allergy? Take the Quiz to Find Out!"
  - **Conclusion:** Summarize the key points and include a call-to-action for readers to explore your hypoallergenic dog food products or contact you for more information.

By improving upon competitors' content, making it more comprehensive, engaging, and up-to-date, you can attract more visitors, retain their interest, and encourage them to take action. This approach not only helps in ranking better in search results but also builds trust and authority with your audience.

Real-World Examples and Case Studies

To illustrate these strategies, let's look at some real-world examples of businesses that have successfully implemented content marketing:

Example 1: Hypoallergenic Dog Food Company

Strategy: A hypoallergenic dog food company creates a series of blog posts addressing common concerns of dog owners, such as "How to Identify Food Allergies in Dogs" and "The Benefits of Hypoallergenic Dog Food." They also produce video testimonials from customers who have seen positive changes in their pets' health. Additionally, they introduced packages and subscription services for their dog food based on competitor analysis.

Outcome: This content attracts dog owners searching for solutions to their pets' food allergies, builds trust in the brand, and increases sales of their hypoallergenic dog food products. The subscription model ensures recurring revenue and improved customer retention.

Example 2: E-commerce Store

Strategy: An online store selling outdoor gear creates detailed buying guides and product reviews. They also produce how-to videos on using their products in different outdoor activities.

Outcome: These comprehensive pieces rank well in search engines and help potential customers make informed purchasing decisions, leading to higher sales.

Example 3: B2B Software Company

Strategy: A B2B software company publishes case studies and whitepapers showcasing how their solutions solve specific business problems. They host webinars and create in-depth blog posts on industry trends.

Outcome: This content positions them as an industry leader and generates leads from businesses looking for similar solutions.

Creating the Right Content

Content marketing is not just about creating content; it's about creating the right content that resonates with your audience and supports your business goals. By understanding your audience, conducting thorough keyword research, developing a strategic content plan, and promoting your content effectively, you can harness the power of content marketing to drive engagement, build trust, and grow your business.

Remember, content is king when it provides real value to your audience. Focus on delivering high-quality, relevant content consistently, and you'll see the positive impact on your digital marketing efforts.

Ready to master content marketing and elevate your business? Let's dive into the strategies and tactics that will help you create content that truly reigns.

Chapter 13

COHESIVE MARKETING STRATEGY

Imagine your digital marketing efforts as a well-oiled machine, where each component works in harmony to drive your business forward. What if you could create a strategy that integrates all the elements you've learned about—understanding your ideal customer, analyzing competitors, conducting keyword research, mastering SEO, and crafting compelling content—into a seamless, effective marketing plan? This is the essence of a cohesive marketing strategy.

In this chapter, we'll explore how to unify your marketing initiatives into a synchronized approach that maximizes your reach and impact. We'll guide you through the process of integrating these elements into a comprehensive strategy that not only enhances your online presence but also drives measurable business growth.

This journey will involve a deep dive into the principles of strategic planning, providing you with actionable insights and practical steps to implement a cohesive marketing strategy.

The Importance of a Cohesive Marketing Strategy

A cohesive marketing strategy ensures that all your marketing efforts are aligned and working towards common goals. It provides a clear roadmap, helping you prioritize tasks, allocate resources efficiently, and measure success accurately. By integrating various marketing elements, you can create a powerful synergy that amplifies the impact of each individual component.

Integrating Key Elements into Your Strategy

To create a cohesive and effective marketing strategy, you need to integrate the following key elements:

- **Understanding Your Ideal Customer**
- **Competitor Analysis**
- **Keyword Research and Analysis**
- **SEO Mastery**
- **Content Marketing**
- **On-Page and Technical SEO**
- **Measuring and Adjusting**

1. Understanding Your Ideal Customer

Your ideal customer is the foundation of your marketing strategy. By creating detailed customer personas, you can tailor your marketing efforts to meet their specific needs and preferences.

- **Review Customer Personas:** Ensure they reflect your current target audience.
- **Tailor Messaging:** Customize your marketing messages to address the pain points, goals, and interests of your personas.
- **Segmentation:** Segment your audience based on demographics, behaviors, and preferences to create more personalized marketing campaigns.

2. Competitor Analysis

Analyzing your competitors provides valuable insights into what works and what doesn't in your industry. Use these insights to identify opportunities, refine your strategies, and differentiate your brand.

- **Monitor Competitors:** Regularly check competitors' websites, content, and social media channels.

- **Identify Gaps:** Look for areas where competitors are lacking and where you can provide more value.
- **Learn from Successes and Failures:** Adapt successful strategies and avoid mistakes made by competitors.

3. Keyword Research and Analysis

Effective keyword research is the backbone of your SEO and content marketing efforts. It helps you understand what your audience is searching for and how to attract them to your site.

- **Conduct Thorough Keyword Research:** Use tools like Google Keyword Planner, Ahrefs, and SEMrush.
- **Focus on Long-Tail Keywords:** Target specific, intent-driven phrases that are more likely to convert.
- **Prioritize Keywords:** Based on search volume, difficulty, and relevance to your business goals.

4. SEO Mastery

SEO is critical for driving organic traffic to your website. It involves optimizing your site to rank higher in search engine results and attract more visitors.

- **On-Page SEO:** Optimize title tags, meta descriptions, header tags, and content for target keywords.
- **Technical SEO:** Ensure your website is technically sound, with fast load times, mobile-friendliness, and proper indexing.
- **Off-Page SEO:** Build high-quality backlinks, engage on social media, and use local SEO tactics.

5. Content Marketing

Content marketing ties all your efforts together by providing valuable, relevant content that engages your audience and supports your business goals.

- **Create a Content Plan:** Develop a content calendar outlining what content you will create and when you will publish it.
- **Focus on Quality:** Produce high-quality, comprehensive content that addresses user intent and provides real value.
- **Promote Your Content:** Use social media, email marketing, influencer collaborations, and paid ads to increase visibility.

6. On-Page and Technical SEO

Optimizing your website's on-page and technical aspects ensures that your content is accessible, user-friendly, and easily discoverable by search engines.

- **Optimize On-Page Elements:** Focus on title tags, meta descriptions, header tags, and internal linking.
- **Improve Technical SEO:** Ensure your site is fast, mobile-friendly, and properly indexed. Use schema markup to enhance search engine understanding.

7. Measuring and Adjusting

Measuring the performance of your marketing efforts is crucial for understanding what works and what doesn't. Use analytics tools to track key metrics and adjust your strategy based on data.

- **Set Clear KPIs:** Define key performance indicators (KPIs) for your marketing goals.
- **Use Analytics Tools:** Track performance using Google Analytics, SEMrush, and social media insights.
- **Adjust Based on Data:** Continuously refine your strategy based on performance data and market trends.

Creating a Cohesive Marketing Plan

To integrate these elements into a cohesive plan, follow these steps:

- **Set Clear Goals**
- **Develop a Unified Strategy**
- **Allocate Resources Efficiently**
- **Implement and Execute**
- **Monitor and Adjust**

Step 1: Set Clear Goals

Start by setting clear, measurable goals for your marketing efforts. These should align with your overall business objectives and provide a clear direction for your strategy.

Example Goals:

- **Increase Website Traffic:** Aim to increase organic traffic by 30% in the next six months.
- **Generate Leads:** Capture 500 new leads per month through content marketing and SEO.
- **Boost Sales:** Increase online sales by 20% in the next quarter.

Step 2: Develop a Unified Strategy

Create a unified strategy that integrates all key elements. This involves planning how each component will work together to achieve your goals.

Example Strategy:

- **Content Marketing:** Create blog posts targeting long-tail keywords identified in your research. Promote these posts through social media and email newsletters.

- **SEO:** Optimize blog posts for on-page SEO and build backlinks through guest blogging and influencer collaborations.
- **Technical SEO:** Ensure your website is fast, mobile-friendly, and properly indexed to support content and SEO efforts.

Step 3: Allocate Resources Efficiently

Allocate your resources, including budget, time, and personnel, to ensure each component of your strategy is adequately supported.

Example Allocation:

- **Content Creation:** Allocate 40% of your budget to content creation, including blog posts, videos, and infographics.
- **SEO and Technical Optimization:** Allocate 30% of your budget to SEO and technical improvements.
- **Promotion:** Allocate 20% of your budget to promoting content through paid ads, social media, and influencer collaborations.
- **Analytics and Adjustment:** Allocate 10% of your budget to monitoring performance and making necessary adjustments.

Step 4: Implement and Execute

Execute your strategy by implementing the planned activities and ensuring all team members are aligned with the goals and tactics.

Example Implementation:

- **Content Calendar:** Follow your content calendar to ensure consistent content production and publication.
- **SEO Practices:** Regularly update your website and content to reflect SEO best practices and new keyword opportunities.

- **Promotion Schedule:** Stick to a promotion schedule that maximizes content visibility and engagement.

Step 5: Monitor and Adjust

Regularly monitor the performance of your marketing efforts and make adjustments as needed. Use analytics tools to track key metrics and identify areas for improvement.

Example Monitoring:

- **Weekly Reviews:** Conduct weekly reviews of website traffic, content performance, and SEO rankings.
- **Monthly Reports:** Create monthly reports to evaluate overall progress and identify trends.
- **Quarterly Adjustments:** Make quarterly adjustments to your strategy based on performance data and market changes.

Chapter 14

MARKETING STRATEGY EXAMPLES

This is an example of a marketing strategy and is intended to be taken as a guide. The numbers, company names, and strategies mentioned herein are fictional and should not be taken as fact.

This example is for illustrative purposes only and should not be interpreted as real data or used as evidence of any specific practices or results.

Ecom B2C - Achieving Your Revenue Goals

Our goal is to help you achieve a 20% growth in gross revenue through a comprehensive and data-driven marketing strategy.

By leveraging targeted customer insights, competitive analysis, effective keyword research, robust technical and on-page SEO, strategic content marketing, and optimized PPC campaigns, we will drive significant improvements in brand visibility, customer engagement, and sales.

This strategy is designed to deliver measurable results and a strong return on investment (ROI), ensuring that every marketing dollar spent contributes to your bottom line.

It is important to note that this strategy is subject to change based on future goals, Google algorithm updates, and market dynamics. Strategies should be fluid and adaptable to ensure continued success in an ever-evolving digital landscape.

Marketing Strategy Example #1

Client: Hypoallergenic Dog Food Company

Objective: Achieve a 20% growth in gross revenue and a strong ROI by increasing brand awareness, driving website traffic, generating leads, nurturing those leads, converting them into customers, enhancing customer retention, and building brand authority.

Understanding Your Ideal Customer

- **Persona: Diligent Dana**
  - **Profile:** A 35-year-old healthcare professional living in an urban area. She is highly interested in pet care, health and wellness, and organic products.
  - **Goals:** To find reliable dog food that won't trigger her dog's allergies.
  - **Pain Points:** Struggles with finding affordable hypoallergenic dog food that her small dog enjoys eating.

Analyzing Your Competitors

- **Example Competitor:** Hypoallergenic Paws
  - **Strengths:** Dedicated page on hypoallergenic dog food explaining its benefits, strong social media presence, high engagement rates, and a robust FAQ section.
  - **Weaknesses:** Lack of detailed guides on transitioning dogs to hypoallergenic food, limited educational content. No content directed at specific breeds or sizes of dogs.

Strategy:

- Develop a more comprehensive hypoallergenic dog food page, including detailed guides and customer testimonials.
- Enhance content with transition guides and educational articles to fill the identified gaps.

Keyword Research and Analysis

- **Seed Keywords:** "dog food allergies," "hypoallergenic dog food," "dog food for sensitive stomachs."
- **Long-Tail Keywords:** "hypoallergenic dog food for small breeds," "natural hypoallergenic dog food recipes."
- **Size-Specific Keywords:**
  - **Small Dogs:** "dog food for small breeds," "small dog food allergies."
  - **Medium Dogs:** "dog food for medium dogs," "medium dog food allergies."
  - **Large Dogs:** "dog food for large breeds," "large dog food allergies."

Technical SEO

- **Mobile Optimization:** Ensure the website is fully responsive on all devices.
- **URL Structure:** Make sure there are clear, descriptive URLs for all pages.
- **XML Sitemaps:** Create and submit sitemaps to improve indexation by search engines.
- **Schema Markup Strategy:**
  - **Product Schema:** Add structured data to product pages to provide search engines with detailed information about your hypoallergenic dog food, including price, availability, and reviews.

- **FAQ Schema:** Implement FAQ schema on relevant pages to enhance search visibility and provide direct answers in search results.
- **Review Schema:** Include review schema to display star ratings and reviews in search results, improving click-through rates.
- **Breadcrumb Schema:** Implement breadcrumb schema to improve site navigation and enhance search appearance.
- **Article Schema:** Use article schema for blog posts to enhance visibility in search results, enabling features like rich snippets.
- **Organization Schema:** Implement organization schema to provide search engines with detailed information about your company, such as name, address, contact details, logo, and social media profiles.

On-Page SEO Best Practices

- **New Service Pages:** Create dedicated service pages for small, medium, and large dogs.
  - **Small Dogs Page:** Information about hypoallergenic food specifically for small breeds.
  - **Medium Dogs Page:** Tailored content for medium-sized dogs.
  - **Large Dogs Page:** Detailed information for large breeds.
- **Title Tags and Meta Descriptions:** Each page should have unique title tags and meta descriptions that include relevant keywords.
  - **Example:** The "Small Dogs" page could have a title tag like "Best Hypoallergenic Dog Food for Small Breeds - [Brand Name]" and a meta description such as "Discover the best hypoallergenic dog food specifically designed for small breeds. Ensure your dog's health with our top-quality products."

- **Headers and Sub headers:** Use H1 for the main title, H2 for major sections, and H3 for subsections. Include keywords naturally in headers.
  - **Example:** On the "Medium Dogs" page, use H2 tags for sections like "Why Choose Hypoallergenic Food for Medium Dogs" and "Top Hypoallergenic Dog Foods for Medium Breeds."
- **Alt Text for Images:** Add descriptive alt text to all images, including relevant keywords to improve accessibility and SEO.
  - **Example:** An image showing a happy medium-sized dog eating could have alt text like "Medium dog eating hypoallergenic dog food."
- **Internal Linking:** Link blog posts and other relevant pages to service pages to improve site navigation and SEO.
  - **Example:** The blog post "How to Transition Your Dog to Hypoallergenic Food" should link to the "Small Dogs," "Medium Dogs," and "Large Dogs" pages with context-specific advice.

Mastering Off-Page SEO

- **Backlink Building:** Develop relationships with pet care blogs and industry influencers to secure high-quality backlinks.
- **Social Media Engagement:** Increase activity on platforms like Facebook, Instagram, and Twitter to drive engagement and traffic.

Google Business Profile

- **Complete Profile:** Ensure all information is accurate and comprehensive.

- **Regular Updates:** Post updates, respond to reviews, and add high-quality photos of products.
- **Customer Interaction:** Encourage satisfied customers to leave positive reviews to boost local search rankings.

Strategic Content Marketing

- **Educational Blog Posts:**
  - "Top 5 Signs Your Dog Has a Food Allergy"
  - "How to Transition Your Dog to Hypoallergenic Food"
  - "Common Dog Food Allergens to Avoid"
  - "DIY Hypoallergenic Dog Food Recipes"

Content Calendar:

- **Content Types:** Blog posts, e-books, case studies, infographics, videos.
  - **Note:** If your business can't afford creating infographics, videos, or other multimedia content (which can be very expensive), you can still have a strong impact on your rankings by writing engaging and value-driven blog posts.
- **Distribution Channels:** Website, social media, email newsletters.
- **Month 1 Plan:**
  - **Week 1:** Blog post on "Top 5 Signs Your Dog Has a Food Allergy"
  - **Week 2:** Video tutorial on "How to Transition Your Dog to Hypoallergenic Food"
  - **Week 3:** Infographic on "Common Dog Food Allergens"
  - **Week 4:** E-book release "The Ultimate Guide to Hypoallergenic Dog Food"
- **Content Plan:**
  - Develop separate content for small, medium, and large dog owners.
  - Tailor blog posts, guides, and videos to address the specific needs of each size category.

Marketing Strategy Example #2

Client: Local Flooring Company

Objective: Achieve local market dominance by increasing leads and conversions, improving customer satisfaction, and enhancing brand reputation.

Understanding Your Ideal Customer

- **Persona: Homeowner Hank**
  - **Profile:** A 45-year-old homeowner living in a suburban area. He is interested in home improvement, quality craftsmanship, and value for money.
  - **Goals:** To find a reliable and professional flooring company to renovate his home.
  - **Pain Points:** Struggles with finding a local flooring company that offers high-quality products at affordable prices and has excellent customer service.

Analyzing Your Competitors

- **Example Competitor:** Not Real Floors Inc.
  - **Strengths:** Strong local presence, high customer ratings, comprehensive portfolio of past projects.
  - **Weaknesses:** Limited online presence, outdated website, and few educational resources for customers.

Strategy

- Develop a more user-friendly website with detailed service pages and an updated portfolio.
- Create educational content to help customers understand flooring options and maintenance tips.

Keyword Research and Analysis

- **Seed Keywords:** "flooring company," "local flooring service," "floor installation."
- **Long-Tail Keywords:** "flooring company near me," "affordable floor installation services," "hardwood floor refinishing"
- **Service-Specific Keywords:**
  - **Hardwood Flooring:** "hardwood floor installation," "hardwood floor refinishing."
  - **Tile Flooring:** "tile floor installation," "tile flooring services."
  - **Carpet Installation:** "carpet installation near me," "best carpet installation service."

Technical SEO

- **Site Speed:** Implement caching and compression techniques to improve load times.
- **Mobile Optimization:** Ensure the website is fully responsive on all devices.
- **URL Structure:** Use clear, descriptive URLs for all pages.
- **XML Sitemaps:** Create and submit sitemaps to improve indexation by search engines.
- **Schema Markup Strategy**
  - **Local Business Schema:** Add structured data to provide search engines with detailed information about your

business, such as address, phone number, and opening hours.

- o **Service Schema:** Implement service schema on service pages to detail the specific flooring services offered.
- o **Review Schema:** Include review schema to display star ratings and reviews in search results, improving click-through rates.
- o **FAQ Schema:** Implement FAQ schema on relevant pages to enhance search visibility and provide direct answers in search results.
- o **Breadcrumb Schema**: Implement breadcrumb schema to improve site navigation and enhance search appearance.

On-Page SEO Best Practices

- **Service Pages:** Create dedicated service pages for different types of flooring services.
  - o **Hardwood Flooring Page:** Information about hardwood floor installation and refinishing.
  - o **Tile Flooring Page:** Details on tile floor installation and maintenance.
  - o **Carpet Installation Page:** Overview of carpet installation services and options.
- **Title Tags and Meta Descriptions:** Each page should have unique title tags and meta descriptions that include relevant keywords.
  - o **Example:** The "Hardwood Flooring" page could have a title tag like "Expert Hardwood Floor Installation - [Company Name]" and a meta description such as "Professional hardwood floor installation and refinishing services. Quality craftsmanship and competitive prices."
- **Headers and Sub headers:** Use H1 for the main title, H2 for major sections, and H3 for subsections. Include keywords naturally in headers.

- o **Example:** On the "Tile Flooring" page, use H2 tags for sections like "Why Choose Tile Flooring" and "Top Tile Flooring Options."
- **Alt Text for Images:** Add descriptive alt text to all images, including relevant keywords to improve accessibility and SEO.
  - o **Example:** An image showing a beautifully tiled floor could have alt text like "High-quality tile flooring installation."
- **Internal Linking:** Link blog posts and other relevant pages to service pages to improve site navigation and SEO.
  - o **Example:** The blog post "How to Choose the Right Tile for Your Home" should link to the "Tile Flooring" page.

Mastering Off-Page SEO

- **Backlink Building:** Develop relationships with home improvement blogs and industry influencers to secure high-quality backlinks.
- **Local Business Listings:** Ensure the business is listed on local directories and review sites like Yelp, Angie's List, and HomeAdvisor.

Google Business Profile

- **Complete Profile:** Ensure all information is accurate and comprehensive.
- **Regular Updates:** Post updates, respond to reviews, and add high-quality photos of recent projects.
- **Customer Interaction:** Encourage satisfied customers to leave positive reviews to boost local search rankings.

Strategic Content Marketing

- **Educational Blog Posts:**
  - "5 Benefits of Hardwood Flooring"
  - "How to Choose the Right Tile for Your Home"
  - "Carpet Installation: What to Expect"
  - "Flooring Maintenance Tips for Homeowners"

Content Calendar:

- **Content Types:** Blog posts, e-books, case studies, infographics, videos.
  - **Note:** If your business can't afford creating infographics, videos, or other multimedia content (which can be very expensive), you can still have a strong impact on your rankings by writing engaging and value-driven blog posts.
- **Distribution Channels:** Website, social media, email newsletters, guest blogs.
- **Month 1 Plan:**
  - **Week 1:** Blog post on "5 Benefits of Hardwood Flooring"
  - **Week 2:** Video tutorial on "How to Choose the Right Tile for Your Home"
  - **Week 3:** Infographic on "Carpet Installation Process"
  - **Week 4:** E-book release "The Ultimate Guide to Flooring Options"

Content Plan:

- Develop separate content for different types of flooring services.
- Tailor blog posts, guides, and videos to address the specific needs of homeowners looking for various flooring solutions.

The Path to Success with Tailored Marketing Strategies

We've explored comprehensive marketing strategies tailored for different types of businesses. From a hypoallergenic dog food company aiming to increase sales and brand authority to a local flooring company seeking to dominate the local market, these examples illustrate how a well-crafted marketing strategy can drive significant growth and success.

Key Takeaways:

- **Understanding Your Audience:** Detailed customer personas help in creating targeted marketing strategies that resonate with your audience's needs and pain points.
- **Competitor Analysis:** Analyzing competitors provides insights into market gaps and opportunities, enabling you to position your business more effectively.
- **Keyword Research and SEO:** Effective keyword research and SEO strategies ensure that your content reaches the right audience, driving organic traffic and improving search engine rankings.
- **Technical and On-Page SEO:** Implementing robust technical and on-page SEO practices enhances website performance, user experience, and search visibility.
- **Content Marketing:** Strategic content marketing, including educational blog posts, videos, and infographics, engages your audience and establishes your brand as an authority in your industry.
- **Adaptability:** Marketing strategies should be fluid and adaptable to changing market conditions, future goals, and algorithm updates.

By following these comprehensive strategies, businesses can achieve their goals, whether it's increasing sales, improving customer satisfaction, or establishing a strong brand presence. Remember, the key to a successful marketing strategy lies in continuous monitoring, analysis, and optimization. Regularly updating your strategies based on performance data and market trends ensures that your business remains competitive and continues to grow.

These examples are not just theoretical frameworks but actionable plans that have been proven to deliver results. Use them as a guide to develop your own tailored marketing strategies that align with your business goals and market dynamics.

By adhering to these principles and customizing the strategies to fit your specific needs, you can pave the way for sustained business growth and success. The journey from understanding your ideal customer to executing a comprehensive marketing plan is integral to achieving your desired outcomes and staying ahead in a competitive landscape.

Chapter 15

BONUS: UNDERSTANDING PPC (PAY-PER-CLICK)

Let's be realistic: Google Ads can be expensive. Can you be successful in Google Ads with a $400 monthly budget? Yes, it's possible, but it heavily depends on various factors like the industry, competition, audience target, product or service offered, and location.

Sometimes, Google Ads or social ads might not be the best fit for your business. This is why I always recommend SEO as a complementary strategy. SEO is a long-term investment that can yield more sustainable and less volatile results compared to running ads.

Is there more to Google Ads or running ads in general than what I've covered? Absolutely. The goal here is to provide you with solid, strategic advice to get you started on the right path.

Introduction to PPC

Pay-Per-Click (PPC) advertising is a digital marketing strategy where advertisers pay a fee each time their ad is clicked. It's a way to buy visits to your site rather than earning them organically. PPC is essential for businesses looking to quickly generate traffic, leads, and sales. When done correctly, PPC can be a powerful tool to reach your target audience, drive conversions, and maximize your return on investment (ROI).

The Benefits of PPC for Business Growth

Immediate Results: Unlike SEO, which can take months to yield results, PPC can drive traffic to your website almost instantly after your campaigns go live.

Targeted Advertising: PPC allows you to target your ads based on demographics, location, keywords, and even specific times of day, ensuring your message reaches the right audience.

Measurable ROI: PPC platforms provide detailed analytics, making it easy to track performance and ROI. This data-driven approach helps in refining strategies for better outcomes.

Flexibility and Control: You have complete control over your budget, bids, and targeting, allowing you to adjust campaigns quickly based on performance and changing business needs.

PPC Budget Management

Determining your PPC budget is a critical step in planning your campaigns. Your budget will influence the volume of traffic you can drive, the competitiveness of your keywords, and the overall effectiveness of your PPC efforts.

How Budgets Work

- **Daily Budget:** This is the amount you're willing to spend each day on a particular campaign. Google will try to get you as many clicks as possible within this budget.
- **Monthly Budget:** Google will pace your spending to ensure you don't exceed your monthly budget, which is essentially your daily budget multiplied by 30.4 (the average number of days in a month).

Deciding How Much to Budget

- **Start Small and Scale:** Begin with a modest budget to test your campaigns. Once you identify what works, you can gradually increase your budget to scale your efforts.
- **Calculate Your CPA:** Determine your cost per acquisition (CPA) target. An acquisition could be a form submission, a phone call, or a sale. This is the amount you're willing to pay to acquire a new customer. Use this to guide your budgeting decisions.
- **Estimate Your CPC:** Use keyword research tools to estimate the cost per click (CPC) for your chosen keywords. Multiply your estimated CPC by the number of clicks you want to receive to determine your daily budget.
- **Set Realistic Goals:** Align your budget with your business goals. If your goal is to generate 100 leads per month and your target CPA is $50, your monthly budget should be $5,000.

Example Calculation:

- **Desired Leads Per Month:** 100
- **Target CPA (Cost Per Acquisition):** $50
- **Monthly Budget:** 100 leads x $50 CPA = $5,000
- **Daily Budget:** $5,000 / 30.4 = Approximately $165

Simple Starting Budget Calculation

Determining a starting budget for your PPC campaigns can be straightforward with a practical example. Let's assume you own a business that operates from 9 AM to 5 PM, Monday to Friday. You decide to run ads only during these hours to align with your business operations. This gives you approximately 20 advertising days per month, excluding holidays.

Here's a step-by-step calculation for setting your initial budget:

- **Determine Clicks Per Day:** Aim for at least 10 clicks per day to ensure sufficient traffic to evaluate your campaign's effectiveness.
- **Estimate Cost Per Click (CPC):** Based on your keyword research, assume your target keywords average $12 per click. Remember, this is an average, so some keywords may cost more while others may cost less.
- **Calculate Daily Budget:** Multiply your estimated CPC by the number of clicks you aim to get per day.
  - **Daily Budget = 10 clicks per day × $12 per click = $120 per day**
- **Calculate Monthly Budget:** Multiply your daily budget by the number of days you plan to run your ads each month.
  - **Monthly Budget = $120 per day × 20 days per month = $2,400 per month**

So, for this example, your initial Google Ads budget would be $2,400 per month.

This calculation ensures that your advertising spend aligns with your business hours and gives you a starting point to assess the performance of your PPC campaigns.

Practical Advice:

- **Monitor and Adjust:** Regularly monitor your campaign performance. If certain keywords or ads are underperforming, adjust your budget allocations accordingly. Or "pause" those ads or keywords.
- **Use Budget Allocation:** Distribute your budget based on campaign performance. Allocate more budget to high-performing campaigns and reduce spending on underperforming ones.
- **Leverage Automated Bidding:** Consider using automated bidding strategies offered by Google Ads to optimize your budget and bids based on performance data. Sometimes it works, sometimes it doesn't.

Understanding Campaign Structure: The Filing Cabinet Analogy

To effectively manage your PPC campaigns, it's essential to understand the structure within Google Ads. Think of it like a filing cabinet:

- **Campaigns:** Campaigns are the drawers of your filing cabinet. Each campaign contains a specific set of goals, budget, and targeting options. For example, you might have separate campaigns for different product lines or geographic regions.
- **Ad Groups:** Within each campaign, you have folders, which are your ad groups. Each ad group contains a set of related keywords and ads. Ad groups help you organize your ads based on themes or product categories.
- **Keywords:** Keywords are the individual files within your folders. These are the terms and phrases that trigger your ads when users search for them.
- **Ads:** Ads are the documents within the files. These are the actual advertisements that users see. Each ad group can contain multiple ads, allowing you to test different messages and determine which performs best.

Creating Effective PPC Campaigns

Before launching a PPC campaign, it's crucial to lay a strong foundation. This involves choosing the right platforms, selecting relevant keywords, and establishing a bidding strategy that aligns with your business goals.

Choosing Platforms

- **Google Ads:** The most popular PPC platform, Google Ads allows you to display ads on Google's search engine results pages (SERPs), YouTube, and other Google properties. It offers extensive reach and sophisticated targeting options. We're going to focus on Google Ads for this lesson.
- **Bing Ads:** Although not as popular as Google Ads, Bing Ads (now Microsoft Advertising) provides access to a different audience segment. It's typically less competitive, which can result in lower CPC (Cost Per Click) and higher ROI.
- **Social Media Platforms:** Platforms like Facebook, Instagram, and LinkedIn also offer PPC advertising options. These are ideal for targeting specific demographics and interests, making them perfect for businesses looking to engage users on social media.

The Reality of Google Reps

Google Ad reps change quarterly and often follow scripts or checklists. While they can provide useful insights, their recommendations, such as expanding to broad keywords or increasing budgets, may not always align with your business goals. Remember, you know your business better than anyone.

Don't let them bully you into strategies that don't fit your objectives. Take their advice with a grain of salt and rely on your deep understanding of your business and strategic objectives. Also, keep in mind that Google, like your business, aims to maximize its revenue. Their suggestions might be driven by this goal as well.

Keyword Selection for PPC

Understanding keyword match types is crucial for effectively managing your PPC campaigns. Each match type controls how closely a keyword needs to match a user's search query to trigger your ad. Google Ads offers three main keyword match types:

- **Broad Match:** Ads may show for searches that include misspellings, synonyms, related searches, and other relevant variations.
  - **Example:** If your keyword is "women's hats," your ad might show for searches like "buy ladies hats," "women's caps," or "women's headwear."
  - **Best Use:** Use broad match to capture a wide audience and gather data on search terms that can inform more precise targeting.
- **Phrase Match:** Ads show for searches that include the exact phrase or close variations of that phrase, with additional words before or after.
  - **Example:** If your keyword is "women's hats," your ad might show for searches like "buy women's hats" or "women's hats for winter," but not for "hats for women."
  - **Best Use:** Use phrase match to reach a more targeted audience while maintaining some flexibility in search queries.
- **Exact Match:** Ads show for searches that match the exact term or close variations of that exact term.
  - *Example:* If your keyword is "women's hats," your ad will only show for searches like "women's hats" or close variants like "women hats."
  - *Best Use:* Use exact match for the highest level of control over who sees your ads. It's ideal for targeting highly specific search terms.
- **Negative Keywords:** Negative keywords are a powerful tool to refine your targeting by excluding specific search terms that are not relevant to your product or service. They can be applied at both the campaign and ad group levels.
  - *Definition:* Negative keywords prevent your ads from showing for searches that include those terms.

- *Example:* If you sell premium women's hats and don't want to attract bargain hunters, you might add "cheap" as a negative keyword.
- *Best Use:* Use negative keywords to filter out irrelevant traffic and ensure your ads are shown to a more qualified audience. This helps in improving your click-through rate (CTR) and conversion rates while reducing wasted spend.

Practical Advice on Using Negative Keywords:

Negative keywords are a powerful tool that can help business owners maximize their online advertising efforts. By strategically using negative keywords, you can ensure your ads are shown to the right audience, thereby maximizing your return on investment (ROI) and reducing wasted ad spend.

This approach is particularly useful when using broad match keywords, which have a wider reach but can sometimes attract irrelevant traffic.

Why Use Negative Keywords?

1. Increased Relevance:

- By excluding irrelevant search terms, your ads will be shown to users who are more likely to be interested in your products or services.
- This leads to higher click-through rates (CTR) and conversion rates, as your ads are tailored to the right audience.

2. Cost Efficiency:

- Negative keywords help in reducing wasted ad spend by preventing clicks from users who are not likely to convert.
- This is crucial for maintaining a healthy return on ad spend (ROAS).

3. Improved Quality Score:

- Google Ads uses Quality Score to measure the relevance and quality of your ads. By targeting the right keywords and excluding irrelevant ones, you can improve your Quality Score.
- A higher Quality Score can lead to lower cost-per-click (CPC) and better ad positions.

How to Use Negative Keywords Effectively

1. Identify Negative Keywords:

- Conduct thorough research to identify terms that are irrelevant to your business.
- Regularly review the search terms report to find keywords that are triggering your ads but not leading to sales or inquiries.

2. Categorize Negative Keywords:

- Group your negative keywords into themes or categories to manage them more effectively.
- For example, if you are an HVAC company, you might have categories like "DIY repairs," "free services," "cheap parts," etc.

3. Apply Negative Keywords at the Campaign and Ad Group Level:

- Depending on the scope of irrelevance, apply negative keywords at the campaign or ad group level.
- Campaign-level negatives are broad and apply to all ad groups within a campaign, while ad group-level negatives are more specific.

4. Use Keyword Match Types for Negative Keywords:

- Just like regular keywords, negative keywords can be broad, phrase, or exact match.
- Use broad match negatives for general exclusions, phrase match to exclude specific phrases, and exact match to exclude specific search terms.

Example Strategy: HVAC Company

Let's consider an HVAC company aiming to attract customers needing professional HVAC services. They want to use broad match keywords to capture a wide audience but need to avoid clicks from users looking for DIY solutions or low-cost parts.

Step 1: Choose Broad Match Keywords

- Broad match keywords might include "HVAC repair," "air conditioning service," and "furnace maintenance."

Step 2: Identify Irrelevant Search Terms

- Analyze search terms that could trigger these broad match keywords but do not align with the company's target audience.
- Examples include terms like "DIY", "free", "cheap", "manual", "parts", "home improvement store", "interview", "job" and "tutorial."

Step 3: Create a Negative Keyword List

- Based on the analysis, create a list of negative keywords such as "DIY HVAC repair", "free air conditioning service", "cheap furnace maintenance", "HVAC repair manual", "HVAC parts", "home improvement store", "HVAC job interview", and "HVAC repair tutorial."

Step 4: Apply Negative Keywords

- Add these negative keywords at both the campaign and ad group levels to ensure they exclude irrelevant traffic across all relevant ads.

Step 5: Monitor and Optimize

- Regularly review the search terms report to identify new irrelevant terms.
- Continuously update the negative keyword list to refine the targeting further.

Tools and Tips for Managing Negative Keywords

1. Use Google Ads Negative Keyword Lists:

- Create and manage negative keyword lists within Google Ads.
- Apply these lists across multiple campaigns to save time and ensure consistency.

2. Leverage Automation Tools:

- Use tools like SEMrush, Ahrefs, or WordStream to automate the identification and application of negative keywords.

- These tools can provide insights into new negative keyword opportunities based on real-time data.

3. Stay Updated with Industry Trends:

- Keep an eye on industry trends and changes in user search behavior.
- Adjust your negative keyword strategy accordingly to remain relevant and efficient.

Tying It All Back to ROI and Business Growth

By meticulously curating and managing negative keyword lists, sometimes consisting of thousands of keywords, you can effectively control your campaigns' reach. This ensures that your ads are shown only to the most relevant searches, ultimately leading to better performance and higher ROI.

For an HVAC company, this means attracting customers who need professional services rather than DIY solutions or people looking to see if you're hiring, resulting in more qualified leads and higher conversion rates. As you refine your targeting, you'll notice a decrease in wasted ad spend and an increase in the efficiency of your marketing budget. This strategic approach not only improves your online presence but also drives tangible growth for your business.

Writing Compelling Ad Copy

Creating compelling ad copy is critical for attracting clicks and conversions. Your ads must stand out and speak directly to your audience's needs.

Headlines, Descriptions, and CTAs

- **Headlines:** Craft clear, concise, and compelling headlines that include your primary keywords. Use emotional triggers or questions to capture attention.
  - **Example:** "Transform Your Home with Elegant Hardwood Flooring" or "Looking for Durable Flooring Solutions? We Have Them!"
- **Descriptions:** Highlight unique selling points (USPs) and benefits. Address potential pain points and how your solution resolves them.
  - **Example:** "Our premium hardwood flooring adds timeless beauty and value to your home. Discover our wide selection and expert installation services today."
- **Calls to Action (CTAs):** Use strong, clear CTAs that tell users what to do next. Use action-oriented language.
- **Ad Extensions:** Utilize ad extensions to provide additional information and improve ad visibility. Popular extensions include sitelink extensions, callout extensions, and structured snippet extensions.

Ad Examples for a Flooring Company

Example 1: Google Search Ad

- **Headline 1:** "Hardwood Flooring Installation"
- **Headline 2:** "Durable & Stylish Flooring Options"
- **Headline 3:** "Free In-Home Consultation"
- **Description 1:** "Enhance your home with our top-quality hardwood flooring. Expert installation. Competitive pricing."
- **Description 2:** "Book a free consultation today to explore our extensive range of flooring solutions."
- **CTA:** "Get a Free Estimate"

Ad Extensions:

- **Sitelink Extension:**
  - "View Flooring Gallery"
  - "Customer Testimonials"
  - "Special Offers"
  - "Contact Us"
- **Callout Extension:** "Free Estimates", "Expert Installers", "10-Year Warranty"
- **Structured Snippet Extension:** "Types: Hardwood, Laminate, Vinyl, Tile"

Example 2: Facebook Ad

- **Image/Video:** High-quality image of a beautifully installed hardwood floor in a modern living room.
- **Headline:** "Transform Your Home with New Hardwood Floors"
- **Description:** "Our expert team provides top-notch hardwood flooring installation. Choose from a wide variety of styles and finishes to match your home decor. Contact us today for a free consultation!"
- **CTA Button:** "Learn More"

Example 3: Instagram Ad

- **Image/Video:** Short video showcasing the before and after transformation of a living room with new hardwood flooring.
- **Headline:** "Discover the Beauty of Hardwood Flooring"
- **Description:** "Upgrade your home with our stunning hardwood flooring options. Our experienced team ensures flawless installation. Click to see more and book your free consultation!"
- **CTA Button:** "Book Now"

Ad Examples for a Window Tinting Company

Example 2: Google Search Ad (Auto Window Tinting)

- **Headline 1:** "Expert Auto Window Tinting Services"
- **Headline 2:** "Protect Your Car with Quality Tint"
- **Headline 3:** "Get a Free Estimate Today"
- **Description 1:** "Reduce heat and glare with our professional auto window tinting. High-quality films, expert installation."
- **Description 2:** "Enhance privacy and protect your car's interior. Contact us now for a free quote."
- **CTA:** "Book Your Appointment Now"

Ad Extensions:

- **Sitelink Extension:**
  - "View Our Work"
  - "Customer Reviews"
  - "Special Discounts"
  - "Contact Us"
- **Callout Extension:** "Free Estimates", "Lifetime Warranty", "UV Protection"
- **Structured Snippet Extension:** "Services: Auto, Residential, Commercial"

Example 2: Facebook Ad (Residential Window Tinting)

- **Image/Video:** High-quality image of a home with professionally tinted windows.
- **Headline:** "Enhance Your Home with Window Tinting"
- **Description:** "Our residential window tinting reduces energy costs, blocks harmful UV rays, and adds privacy. Get a free consultation today!"
- **CTA Button:** "Get a Free Quote"

Example 3: Instagram Ad (Commercial Window Tinting)

- **Image/Video:** Short video showcasing the installation of window tinting in a commercial office space.
- **Headline:** "Professional Commercial Window Tinting"

- **Description:** "Improve your office environment with our commercial window tinting services. Reduce glare, enhance privacy, and increase energy efficiency. Contact us today!"
- **CTA Button:** "Learn More"

By tailoring your ad copy to speak directly to your target audience's needs and showcasing your unique selling points, you'll create compelling ads that drive clicks and conversions for your window tinting company, whether for auto, residential, or commercial services.

Landing Page Optimization

Driving traffic is only part of the equation. Converting that traffic requires optimized landing pages.

Relevance and User Experience

- **Match Ad Copy with Landing Page Content:** Ensure that your landing page content aligns with your ad copy. If your ad promotes a free consultation, the landing page should prominently feature this offer.
- **Clear and Compelling Headlines:** Use clear headlines that immediately convey the value proposition and match the promise made in the ad.
- **Strong Visuals:** Use high-quality images and videos to engage visitors and reinforce your message. Visuals should be relevant and help illustrate benefits.
- **Simple and Intuitive Layout:** Design your landing page with a clean and intuitive layout. Avoid clutter and ensure the most important information and CTA are above the fold.
- **Trust Signals:** Include testimonials, reviews, certifications, and security badges to build credibility and reassure visitors.

- **Fast Load Times:** Ensure your landing page loads quickly. Slow load times can increase bounce rates and negatively impact your Quality Score.

Example: A landing page for a financial planning service should have a clear headline like "Achieve Your Financial Goals" with a strong visual of a happy family, trust signals like client testimonials, and a clear CTA like "Get Your Free Financial Consultation."

Analyzing and Optimizing PPC Campaigns

PPC requires continuous monitoring and optimization to ensure sustained performance and ROI.

Key Metrics and Adjustments

- **Click-Through Rate (CTR):** Monitor CTR to assess ad performance. Low CTR may indicate the need for better ad copy or more relevant keywords.
- **Conversion Rate:** Track conversion rate to measure how many visitors complete desired actions. Optimize landing pages and CTAs to improve this metric.
- **Cost Per Click (CPC) and Cost Per Acquisition (CPA):** Monitor CPC and CPA to ensure you're getting the best value for your budget. Adjust bids and targeting to lower costs and increase efficiency.
- **Quality Score:** Google assigns a Quality Score to each keyword based on ad relevance, landing page experience, and expected CTR. A higher Quality Score can lower your CPC and improve ad positioning.
- **A/B Testing:** Continuously run A/B tests on ad copy, headlines, landing pages, and CTAs. Use results to refine your campaigns and improve performance.

- **Analytics and Reporting:** Use Google Analytics and PPC platform analytics to gather insights and track performance. Regularly review reports and adjust your strategy based on data.

Example: For a seasonal campaign, a retail business could monitor the performance of various holiday-themed ads and landing pages, adjusting bids and testing different CTAs like "Shop Holiday Deals" versus "Limited Time Offers."

Advanced PPC Strategies

Once you've mastered the basics, you can explore advanced PPC strategies to further enhance your campaigns. These strategies may involve more sophisticated targeting, bidding techniques, and integrating other marketing channels.

- **Remarketing:** Target users who have previously visited your website with tailored ads to re-engage them and encourage conversions.
- **Dynamic Search Ads:** Automatically generate ads based on the content of your website, ensuring your ads stay relevant and up-to-date.
- **Custom Audiences:** Create highly targeted ads by combining multiple data sources such as website visits, customer lists, and user behavior.
- **Ad Scheduling:** Optimize your ad spend by scheduling your ads to run during peak hours when your target audience is most active.
- **Geo-Targeting:** Focus your ads on specific geographic areas to reach local customers more effectively.
- **Device Targeting:** Adjust your bids based on the device users are using, such as mobile, desktop, or tablet, to optimize performance.

Learning Phase and Consistency in Google Ads

Google Ads campaigns go through a learning phase each time a significant change is made. During this period, Google's algorithm collects data to understand how your campaign performs and how best to optimize it for future performance.

Learning Phase

- **What It Is:** The learning phase is when Google Ads adjusts to changes in your campaign, such as new keywords, ads, or bid adjustments. During this time, performance may fluctuate as the system gathers data.
- **Duration:** Typically, the learning phase lasts about 7 days, but this can vary depending on the volume of data Google collects.
- **Implications:** Be cautious about making frequent changes to your campaign. Each significant adjustment resets the learning phase, which can delay achieving optimal performance.

Consistency and Reliability

Google values consistency and reliability in your ad campaigns. Regularly pausing and restarting your ads can negatively impact your performance and ranking. Here's why maintaining a consistent presence is crucial:

- **Ad Ranking:** Consistent campaigns are more likely to achieve higher ad rankings and lower costs per click (CPC). Google rewards stable campaigns with better positioning.
- **Budget Efficiency:** Frequent starts and stops can lead to inefficient budget use and increased costs. Google's algorithm may not fully optimize your bids and targeting if your campaign is not running consistently.
- **Competitor Advantage:** Reliable campaigns often outmatch competitors who frequently change or pause their ads.

Consistency builds credibility and trust with both Google and your target audience.

Practical Advice

- **Avoid Frequent Changes:** Make changes to your campaign in phases rather than all at once. This helps minimize the impact on the learning phase.
- **Monitor Performance:** Regularly review your campaign performance but avoid the temptation to make constant tweaks. Allow your campaigns time to stabilize and gather sufficient data for accurate analysis.
- **Optimize Based on Data:** Use data-driven insights to make informed adjustments. This approach ensures changes are meaningful and likely to improve performance.

Viewing Your Ads

Repeatedly searching for your ads without clicking on them can send the wrong signals to Google. The system might interpret your lack of interaction as disinterest, leading to your ads being shown less frequently to you.

However, this does not mean your ads are being shown less frequently to everyone else. It's best to avoid searching for your ads directly, as this can create misleading impressions about their visibility and performance.

Tips to Avoid This Pitfall

- **Ad Preview Tool:** Use the Google Ads Ad Preview and Diagnosis tool to check your ads' visibility without affecting performance metrics.

- **Focus on Metrics:** Instead of manually searching for your ads, rely on Google Ads metrics and reports to understand your campaign's performance.
- **Trust the System:** Google's algorithm is designed to show your ads to the most relevant audience. Trust the data and focus on optimization rather than constantly checking for ad visibility.

Integrating PPC with SEO

SEO and PPC are often seen as separate strategies, but they can complement each other effectively when integrated. Here's how to leverage both for maximum impact:

- **Keyword Data Sharing:** Use PPC keyword data to inform your SEO strategy and vice versa. High-performing PPC keywords can guide your organic keyword targeting.
- **SERP Domination:** By appearing in both paid and organic search results, you can increase your visibility and authority on search engine results pages (SERPs).
- **Remarketing:** Use SEO-driven traffic for remarketing campaigns in PPC. Visitors who come through organic search can be retargeted with PPC ads to drive conversions.
- **Landing Page Optimization:** Optimize your landing pages for both PPC and SEO to improve quality scores and organic rankings simultaneously.
- **A/B Testing Insights:** Use A/B testing results from PPC campaigns to inform your SEO content strategies. Successful ad copy can inspire meta descriptions and page titles.
- **Budget Allocation:** Balance your budget between SEO and PPC based on performance metrics. Shift funds to the strategy delivering the best ROI while maintaining a long-term investment in SEO.

By integrating SEO and PPC, you can create a more comprehensive and effective digital marketing strategy that leverages the strengths of both approaches to drive traffic, increase conversions, and maximize ROI.

Leveraging PPC for Sustainable Success

As we conclude this comprehensive guide on PPC, it's important to remember that successful digital marketing is an ongoing journey, not a one-time effort. The strategies, tips, and insights shared in this chapter are designed to help you lay a strong foundation and make informed decisions as you navigate the complexities of PPC advertising.

Digital marketing, especially PPC, is a dynamic field that continuously evolves. Staying up-to-date with the latest trends, tools, and best practices is crucial for maintaining and enhancing your campaign's performance. Always be prepared to adapt and optimize your strategies based on data-driven insights and emerging opportunities.

While PPC can deliver immediate results, integrating it with long-term strategies like SEO will ensure sustainable growth and a robust online presence. Remember, the ultimate goal is to create a balanced and cohesive digital marketing strategy that aligns with your business objectives and meets the needs of your target audience.

Chapter 16

EPILOGUE: YOUR JOURNEY TO MARKETING MASTERY

Congratulations on reaching the end of this comprehensive guide to crafting effective marketing strategies. Throughout this book, we've covered a wide range of topics essential for any business looking to thrive in today's competitive landscape. From understanding your ideal customer and analyzing your competitors to understanding SEO and PPC, each chapter has been designed to equip you with the knowledge and tools necessary to build a robust marketing plan.

As you embark on your journey to implement the strategies and insights shared in this book, remember that marketing is both an art and a science. It requires creativity, analytical thinking, and a willingness to experiment and learn from data. The road to marketing mastery is continuous, with new trends and technologies always on the horizon.

Stay curious, keep learning, and be proactive in refining your strategies. Engage with your audience authentically, and don't shy away from trying new approaches. The marketing landscape will continue to evolve, and your ability to adapt will be a key determinant of your success.

Thank you for choosing this book as your guide. We hope it serves as a valuable resource and inspires you to achieve remarkable results in your marketing endeavors. Here's to your continued growth and success!

Chapter 17

GLOSSARY

Algorithm: A set of rules or a procedure for solving a problem in a finite number of steps. In SEO, algorithms are used by search engines to determine the relevance and ranking of web pages.

Alt Text (Alternative Text): Descriptive text used in HTML code to describe images. It helps search engines understand what an image is about and improves accessibility for visually impaired users.

Backlink: A link from one website to another. Backlinks are important for SEO because they signal to search engines that your content is valuable and authoritative.

Bounce Rate: The percentage of visitors who leave a website after viewing only one page. A high bounce rate may indicate that visitors are not finding what they are looking for.

Conversion: A desired action taken by a user on a website, such as making a purchase, filling out a form, or signing up for a newsletter.

Conversion Rate: The percentage of users who complete a desired action (conversion) out of the total number of visitors. It is calculated as (Conversions/Total Visitors) * 100.

Cost Per Acquisition (CPA): The cost associated with acquiring a new customer through advertising or marketing efforts. It is calculated as the total cost of marketing divided by the number of conversions.

Cost Per Click (CPC): The amount an advertiser pays each time their ad is clicked in a pay-per-click (PPC) campaign. It is calculated as the total cost of clicks divided by the number of clicks.

Cost Per Conversion: The total cost of generating a conversion. It is calculated by dividing the total cost of a marketing campaign by the number of conversions it generated.

Crawl: The process by which search engines discover and index new or updated web pages. Search engine bots (crawlers) follow links from page to page to collect data about websites.

Customer Lifetime Value (CLV): The total revenue a business can expect from a single customer account throughout its relationship with the customer. It helps businesses understand the long-term value of acquiring and retaining customers.

Domain Authority (DA): A metric developed by Moz that predicts how well a website will rank on search engine result pages (SERPs). It is based on various factors, including the number and quality of backlinks.

Google Analytics: A free tool provided by Google that tracks and reports website traffic. It provides insights into how users find and interact with your website.

Google Business Profile: A free tool that allows businesses to manage their online presence on Google, including Search and Maps. It helps businesses attract local customers and provides important information such as business hours and contact details.

Google Search Console: A free tool provided by Google that helps you monitor, maintain, and troubleshoot your website's presence in Google Search results. It provides insights into how Google views your site and helps you optimize its performance.

Google Tag Manager: A free tool from Google that allows you to manage and deploy marketing tags (snippets of code or tracking pixels) on your website without having to modify the code directly. It simplifies the process of adding and updating tags for analytics, remarketing, and other tracking purposes.

Keyword: A word or phrase that users enter into search engines when looking for information. Keywords are essential for SEO as they help search engines understand the content of your web pages.

Keyword Research: The process of identifying and analyzing the keywords that are relevant to your business and that users are searching for. Effective keyword research helps optimize content to rank higher in search results.

Meta Description: A brief summary of a web page's content that appears in search engine results below the page title. Meta descriptions can influence click-through rates but are not a direct ranking factor.

On-Page SEO: The practice of optimizing individual web pages to rank higher and earn more relevant traffic in search engines. It involves optimizing content, HTML code, and internal links.

Off-Page SEO: The practice of improving a website's visibility and authority through activities outside the website itself, such as building backlinks, social media marketing, and influencer outreach.

Page Rank: A metric used by Google to measure the importance of a web page based on the quantity and quality of backlinks. Higher page rank indicates greater authority and relevance.

Pay-Per-Click (PPC): An online advertising model where advertisers pay a fee each time their ad is clicked. Google Ads is a popular PPC platform.

People Also Ask: A feature in Google search results that displays a list of questions related to the user's query. Each question can be expanded to reveal a brief answer and a link to the source page.

Persona: A detailed description of a fictional user or customer that represents a segment of your target audience. Personas help businesses understand their customers' needs, behaviors, and preferences.

Query: The word or phrase that a user types into a search engine to find information. Queries are essential for search engines to understand user intent and provide relevant search results.

Rich Results: Enhanced search results that go beyond the standard blue link, often featuring images, carousels, reviews, or other interactive elements. Rich results are powered by structured data and schema markup.

Schema Markup: A type of microdata that helps search engines understand the content of your web pages more effectively. Implementing schema markup can enhance your site's appearance in search results with rich snippets.

Search Engine Optimization (SEO): The process of optimizing a website to improve its visibility and ranking in search engine results. SEO encompasses various strategies and techniques, including keyword research, on-page and off-page SEO, and technical SEO.

Search Engine Results Page (SERP): The page displayed by a search engine in response to a user's query. SERPs typically include organic search results, paid ads, and other features like featured snippets and knowledge panels.

Technical SEO: The practice of optimizing a website's technical aspects to improve its crawlability, indexability, and overall performance. This includes optimizing site speed, mobile-friendliness, and URL structure.

Title Tag: An HTML element that specifies the title of a web page. Title tags are displayed in search engine results as the clickable headline and are important for both SEO and user experience.

Traffic: The number of visitors who visit a website. Traffic can come from various sources, including organic search, paid ads, social media, and direct visits.

User: An individual who visits and interacts with your website. Understanding user behavior and preferences is crucial for optimizing the user experience (UX).

User Experience (UX): The overall experience of a user when interacting with a website or application. Good UX is essential for keeping visitors engaged and reducing bounce rates.

XML Sitemap: A file that lists all the pages on a website, helping search engines crawl and index the site more effectively. XML sitemaps are especially useful for large websites with many pages.

www.ingramcontent.com/pod-product-compliance
Lightning Source LLC
Chambersburg PA
CBHW071235210326
41597CB00016B/2066